ENVIRONMENTAL SCIENCE, ENGINEERING AND TECHNOLOGY

COAL-FIRED POWER PLANTS AND CARBON DIOXIDE ISSUES

Environmental Science, Engineering and Technology

Additional books in this series can be found on Nova's website under the Series tab.

Additional E-books in this series can be found on Nova's website under the E-books tab.

Energy Policies, Politics and Prices

Additional books in this series can be found on Nova's website under the Series tab.

Additional E-books in this series can be found on Nova's website under the E-books tab.

ENVIRONMENTAL SCIENCE, ENGINEERING AND TECHNOLOGY

COAL-FIRED POWER PLANTS AND CARBON DIOXIDE ISSUES

MATTHEW W. SHUESTER
EDITOR

Nova Science Publishers, Inc.
New York

For permission to use material from this book please contact us:
Telephone 631-231-7269; Fax 631-231-8175
Web Site: http://www.novapublishers.com

NOTICE TO THE READER

LIBRARY OF CONGRESS CATALOGING-IN-PUBLICATION DATA
Coal-fired power plants and carbon dioxide issues / editors, Matthew W. Shuester.
p. cm.
Includes index.
ISBN 978-1-61122-054-4 (hardcover)
1. Clean coal technologies. 2. Coal-fired power plants--Environmental aspects. 3. Carbon dioxide mitigation. I. Shuester, Matthew W.
TK1051.C54 2010
363.738'746--dc22
2010038636

Published by Nova Science Publishers, Inc. † New York

CONTENTS

PREFACE

Any comprehensive approach to substantially reduce greenhouse gases must address the world's dependency on coal for one-quarter of its energy demand, including almost half of its electricity demand. To maintain coal in the world's energy mix in a carbon-constrained future would require development of a technology to capture and store its carbon dioxide emissions. This situation suggests to some that any greenhouse gas reduction program be delayed until such carbon capture technology has been demonstrated. However, technological innovation and the demands of a carbon control regime are interlinked; a technology policy is no substitute for environmental policy and should be developed in concert with it. This book explores developing a means to control coal-derived greenhouse gas emissions which is an imperative if serious reductions in worldwide emissions are to occur in the foreseeable future.

Chapter 1- Any comprehensive approach to substantially reduce greenhouse gases must address the world's dependency on coal for one-quarter of its energy demand, including almost half of its electricity demand. To maintain coal in the world's energy mix in a carbon-constrained future would require development of a technology to capture and store its carbon dioxide emissions. This situation suggests to some that any greenhouse gas reduction program be delayed until such carbon capture technology has been demonstrated. However, technological innovation and the demands of a carbon control regime are interlinked; a technology policy is no substitute for environmental policy and should be developed in concert with it.

Much of the debate about developing and commercializing carbon capture technology has focused on the role of research, development, and deployment (technology-push mechanisms). However, for technology to be fully commercialized, it must also meet a market demand—a demand created either through a price mechanism or a regulatory requirement (demand-pull mechanisms). Any conceivable carbon capture technology for coal-fired power plants will increase the cost of electricity generation from affected plants because of efficiency losses. Therefore, few companies are likely to install such technology until they are required to, either by regulation or by a carbon price. Regulated industries may find their regulators reluctant to accept the risks and cost of installing technology that is not required.

Chapter 2- Coal power plants generate about half of the United States' electricity and are expected to remain a key energy source. Coal power plants also account for about one-third of the nation's emissions of carbon dioxide (CO_2), the primary greenhouse gas that experts believe contributes to climate change. Current regulatory efforts and proposed legislation that seek to reduce CO_2 emissions could affect coal power plants. Two key technologies show potential for reducing CO_2 emissions: (1) carbon capture and storage (CCS), which involves capturing and storing CO_2 in geologic formations, and (2) plant efficiency improvements that allow plants to use less coal.

The Department of Energy (DOE) plays a key role in accelerating the commercial availability of these technologies and devoted more than $600 million to them in fiscal year 2009. Congress asked GAO to examine (1) the maturity of these technologies; (2) their potential for commercial use, and any challenges to their use; and (3) possible implications of deploying these technologies. To conduct this work, GAO reviewed reports and interviewed stakeholders with expertise in coal technologies.

Chapter 3- Coal-fired power plants generate about one-half of the nation's electricity and about one-third of its carbon dioxide (CO_2) emissions, which contribute to climate change. In 2003, the Department of Energy (DOE) initiated FutureGen—a commercial-scale, coal-fired power plant to incorporate integrated gasification combined cycle (IGCC), an advanced generating technology, with carbon capture and storage (CCS). The plant was to capture and store underground about 90 percent of its CO_2 emissions. DOE's cost share was 74 percent, and industry partners agreed to fund the rest. Concerned about escalating costs, DOE restructured FutureGen. GAO was asked to examine (1) the original and restructured programs' goals, (2) similarities and differences between the new FutureGen and other DOE CCS programs, and (3) if the restructuring decision was based on sufficient information.

GAO reviewed best practices for making programmatic decisions, FutureGen plans and budgets, and documents on the restructuring of FutureGen. GAO contacted DOE, industry partners, and experts.

In: Coal-Fired Power Plants and Carbon… ISBN: 978-1-61122-054-4
Editor: Matthew W. Shuester

Chapter 1

CAPTURING CO_2 FROM COAL-FIRED POWER PLANTS: CHALLENGES FOR A COMPREHENSIVE STRATEGY

Larry Parker and Peter Folger

SUMMARY

Any comprehensive approach to substantially reduce greenhouse gases must address the world's dependency on coal for one-quarter of its energy demand, including almost half of its electricity demand. To maintain coal in the world's energy mix in a carbon-constrained future would require development of a technology to capture and store its carbon dioxide emissions. This situation suggests to some that any greenhouse gas reduction program be delayed until such carbon capture technology has been demonstrated. However, technological innovation and the demands of a carbon control regime are interlinked; a technology policy is no substitute for environmental policy and should be developed in concert with it.

Much of the debate about developing and commercializing carbon capture technology has focused on the role of research, development, and deployment (technology-push mechanisms). However, for technology to be fully commercialized, it must also meet a market demand—a demand created either through a price mechanism or a regulatory requirement (demand-pull mechanisms). Any conceivable carbon capture technology for coal-fired power plants will increase the cost of electricity generation from affected plants because of efficiency losses. Therefore, few companies are likely to install such technology until they are required to, either by regulation or by a carbon price. Regulated industries may find their regulators reluctant to accept the risks and cost of installing technology that is not required.

The Department of Energy (DOE) has invested millions of dollars since 1997 in carbon capture technology research and development (R&D), and the question remains whether it has been too much, too little, or about the right amount. In addition to appropriating funds each year for the DOE program, Congress supported R&D investment through provisions for loan guarantees and tax credits. Congress also authorized a significant expansion of carbon capture and sequestration (CCS) spending at DOE in the Energy Independence and Security Act of 2007. Funding for carbon capture technology has increased substantially as a result of enactment of the American Recovery and Reinvestment Act of 2009.

Legislation introduced in the 111th and 110th Congresses invokes the symbolism of the Manhattan project of the 1940s and the Apollo program of the 1 960s to frame proposals for large-scale energy policy initiatives that include developing CCS technology. However, commercialization of technology and integration of technology into the private market were not goals of either the Manhattan project or Apollo program.

Finally, it should be noted that the status quo for coal with respect to climate change legislation isn't necessarily the same as "business as usual." The financial markets and regulatory authorities appear to be hedging their bets on the outcomes of any federal legislation with respect to greenhouse gas reductions, and becoming increasingly unwilling to accept the risk of a coal-fired power plant with or without carbon capture capacity. The lack of a regulatory scheme presents numerous risks to any research and development effort designed to develop carbon capture technology. Ultimately, it also presents a risk to the future of coal.

Introduction: Coal and Greenhouse Gas Emissions

The world meets 25% of its primary energy demand with coal, a number projected to increase steadily over the next 25 years. Overall, coal is responsible for about 20% of global greenhouse gas emissions.[1] With respect to carbon dioxide (CO_2), the most prevalent greenhouse gas, coal combustion was responsible for 41% of the world's CO_2 emissions in 2005 (11 billion metric tons).[2]

Coal is particularly important for electricity supply. In 2005, coal was responsible for about 46% of the world's power generation, including 50% of the electricity generated in the United States, 89% of the electricity generated in China, and 81% of the electricity generated in India.[3] Coal- fired power generation is estimated to increase 2.3% annually through 2030, with resulting CO_2 emissions estimated to increase from 7.9 billion metric tons per year to 13.9 billion metric tons per year. For example, during 2006, it is estimated that China added over 90 gigawatts (GW) of new coal-fired generating capacity, potentially adding an additional 500 million metric tons of CO_2 to the atmosphere annually.[4]

Developing a means to control coal-derived greenhouse gas emissions is an imperative if serious reductions in worldwide emissions are to occur in the foreseeable future. Developing technology to accomplish this task in an environmentally, economically, and operationally acceptable manner has been an ongoing interest of the federal government and energy companies for a decade, but no commercial system to capture and store these emissions is currently available for large-scale coal-fired power plants.

Arguably the most economic and technologically challenging part of the carbon capture and sequestration (CCS) equation is capturing the carbon and preparing it for transport and storage.[5] Depending on site-specific conditions, the capture component of a CCS system can be the dominant cost-variable, and the component that could be improved most dramatically by further technological advancement. As indicated in Table 1, capture costs could be 5-10 times the cost of storage. Breakthrough technologies that substantially reduce the cost of capturing CO_2 from existing or new power plants, for example by 50% or more, would immediately reshape the economics of CCS. Moreover, technological breakthroughs would change the economics of CCS irrespective of a regulatory framework that emerges and governs how CO_2 is transported away from the power plant and sequestered underground.

In contrast, the cost of transporting CO_2 and sequestering it underground is likely less dependent on technological breakthroughs than on other factors, such as:

- the costs of construction materials and labor (in the case of pipelines for CO_2 transport);

- the degree of geologic characterization required to permit sequestration;
- the requirements for measuring, monitoring, and verifying permanent CO_2 storage;
- the costs of acquiring surface and subsurface rights to store CO_2;
- costs of insurance and long-term liability; and
- other variables driving the cost of transportation and sequestration.[6]

That is not to say that the transportation and storage components of CCS are independent of cost and timing. Depending on the degree of public acceptance of a large-scale CCS enterprise, the transportation and sequestration costs could be very large, and it may take years to reach agreement on the regulatory framework that would guide long-term CO_2 sequestration. But the variables driving cost and timing for the transportation and storage of CO_2 are less amenable to technological solution.

This report examines the current effort to develop technology that would capture CO_2. First, the report outlines the current status of carbon capture technology. Second, the report examines the role of government in developing that technology, both in terms of creating a market for carbon capture technology and encouraging development of the technology. Finally, the report concludes with a discussion of implications of capture technology for climate change legislation.

Table 1. Expected Costs of CCS Technology Elements

CCS Element	\$/Metric Ton of CO_2
Capture	\$40-\$80
Storage	\$3-\$8
Monitoring and Verification	\$0.2-\$ 1.0

Source: S. Julio Friedmann, Carbon Capture and Sequestration As a Major Greenhouse Gas Abatement Option (November 2007), p. 11.

Note: Capture and storage costs are very site-specific. These estimates reflect the magnitude of difference between capture and storage costs; actual site-specific costs could vary substantially from these estimates. Estimates do not include any transportation costs.

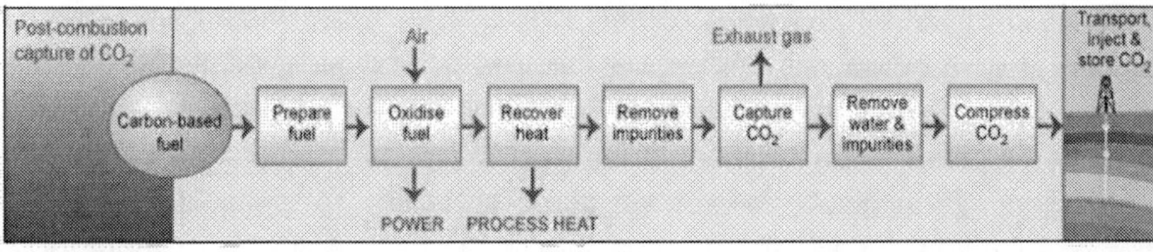

Source: Scottish Centre for Carbon Storage. Figure available at http://www.geos.ed .ac.uk/sccs/capture/ postcombustion.html.

Figure 1. Simplified Illustration of Post-Combustion CO_2 Capture

BACKGROUND: WHAT IS CARBON CAPTURE TECHNOLOGY AND WHAT IS ITS STATUS?

Major reductions in coal-fired CO_2 emissions would require either pre-combustion, combustion modification, or post-combustion devices to capture the CO_2. Because there is currently over 300 GW of coal-fired electric generating capacity in the United States and about 600 GW in China, a retrofittable post-combustion capture device could have a substantial market, depending on the specifics of any climate change program. The following discussion provides a brief summary of technology under development that may be available in the near term. It is not an exhaustive survey of the technological initiatives currently underway in this area, but illustrative of the range of activity. Funding for current government research and development activities to improve these technologies and move them to commercialization are discussed later.

Post-Combustion CO_2 Capture

Post-combustion CO_2 capture involves treating the burner exhaust gases immediately before they enter the stack. The advantage of this approach is that it would allow retrofit at existing facilities that can accommodate the necessary capturing hardware and ancillary equipment. In this sense, it is like retrofitting post-combustion sulfur dioxide (SO2), nitrogen oxides (NOx), or particulate control on an existing facility. A simplified illustration of this process is provided in **Figure 1**.

Post-combustion processes capture the CO_2 from the exhaust gas through the use of distillation, membranes, or absorption (physical or chemical). The most widely used capture technology is the chemical absorption process using amines (typically monoethanolamine (MEA)) available for industrial applications. Pilot-plant research on using ammonia (also an amine) as the chemical solvent is currently underway with demonstration plants announced. These approaches to carbon capture are discussed below. Numerous other solvent-based post-combustion processes are in the bench-scale stage.[7]

Monoethanolamine (MEA)

The MEA CO_2 carbon capture process is the most proven and tested capture process available. The basic design (common to most solvent-based processes) involves passing the exhaust gasesthrough an absorber where the MEA interacts with the CO_2 and absorbs it. The now CO_2-rich MEA is then pumped to a stripper (also called a regenerator) which uses steam to separate the CO_2 from the MEA. Water is removed from the resulting CO_2, which is compressed while the regenerated MEA is purged of any contaminants (such as ammonium sulfate) and recirculated back to the absorber. The process can be optimized to remove 90-95% of the CO_2 from the flue gas.[8]

Although proven on an industrial scale, it has not been applied to the typically larger volumes of flue gas streams created by coal-fired power plants. The technology has three main drawbacks that would make current use on a coal-fired power plant quite costly. First is the need to divert steam away from its primary use—generating electricity—to be used instead for stripping CO_2 from MEA. A second related problem is the energy required to compress the CO_2 after it's captured—needed for transport through pipelines—which lowers overall power plant efficiency and increases generating costs. A study by the Massachusetts Institute of Technology (MIT) estimated the efficiency losses from the installation of MEA from 25%-28% for new construction and 36%-42% for retrofit on an existing plant.[9] This loss of efficiency comes in addition to the necessary capital and operations and maintenance cost of the equipment and reagents. For new construction, the increase in electricity generating cost on a levelized basis would be 60%-70%, depending on the boiler technology. [10] In the case of retrofit plants where the capital costs were fully amortized, the MEA capture process would increase generating costs on a levelized basis by about 220%-250%.1 [11]

A third drawback is degradation of the amine through either overheating (over 205 degrees Fahrenheit [F]) in the absorber or through oxidation from oxygen introduced in the wash water, chemical slurry, or flue gas that reacts with the MEA. For example, residual SO_2 in the flue gas will react with the MEA to form ammonium sulfate that must be purged from the system. [12] This could be a serious problem for existing plants that do not have highly efficient flue gas desulfurization (FGD) or selective catalytic reduction (SCR) devices (or none), requiring either upgrading of existing FGD and SCR equipment, or installation of them in addition to the MEA process.

MEA-Based Solvents (Fluor)

One approach to addressing the shortcomings of MEA identified above is to use an MEA-based formulation of different amines. The goal of the formulation is improve the reaction rates and CO_2 carrying capacity of the solvent under typical power plant conditions. One such proprietary amine-based technology has been developed by Fluor (called Economine FG Plus).[13] Fluor statesthat its MEA-based solvent is specially formulated to recover CO_2 from low pressure, oxygen- containing streams, such as boiler stack gases found in conventional coal-fired power plants. The treatment process is similar to generic gas treating processes, using simple, time-tested gas treating equipment. To address contamination of the solvent, the Fluor technology focuses on cleaning up impurities in the flue gas (e.g., SO_2, NOx, hydrochloric acid, and hydrofluoric acid) in order to inhibit the formulation of ammonium sulfate and other heat-stable salts.[14] Thus the process requires installation of modern SO_2 (with additional enhancements), NOx, Hg, and particulate controls upstream of the carbon capture technology to address the impurities. Solvent temperature issues are dealt with through the use of an intercooler system within the absorber. Finally, other enhancements in the process reduce the process energy consumption.

Fluor 's technology has significant experience on a small commercial scale, including the only CO_2 recovery unit that has operated on the flue gas of a natural gas-fired turbine power plant under typical power plant conditions. Between 1991 and 2005, the Fluor facility at the Bellingham, MA, power plant captured 360 shorts tons of CO_2 daily for use in the food and beverage industry.[15] In September 2009, NRG Energy and Fluor announced that Economine FG Plus would be installed on a 60 megawatt (MW) slipstream[16] at NRG's WA Parish plant (Unit 7) in Texas, with a plan to be operational by 2013. The captured CO_2 would be sold or used for enhanced oil recovery (EOR). DOE will provide up to $154 million toward the project.

Mixed Amine Solvents (HTC Purenergy)

Other combinations of amines and other chemicals have been developed to increase the efficiency and reduce the energy requirements of capture processes. Using blends of water, amines, and other chemicals developed by the University of Regina's International Test Centre for CO_2 Capture, the Purenergy CCS Capture System is designed to be modular, shop-built, and transported to the site.[17]Currently a module is designed to handle the flow of a 50 MW facility with multi-unit train installation possible.

In December 2009, a commercial demonstration for Purenergy CCS was announced by Basin Electric and HTC Pure Energy, using a 20 MW slipstream at Basin Electric's Antelope Valley Station in North Dakota. Commercial operation is planned for 2012. The CO_2 produced will be transported via an existing CO_2 pipeline to Canada for use in EOR.

Chilled Ammonia (Alstom)

An approach to mitigating the oxidation problem identified above is to use an ammonia-based solvent rather than MEA. Ammonia is an amine that absorbs CO_2 at a slower rate than MEA. In achilled ammonia process, the flue gas temperature is reduced from about 130 degrees F to about 35-60 degrees F. This lower temperature has two benefits: (1) it condenses the residual water in the flue gas, which minimizes the volume of flue gas entering the absorber; and (2) it causespollutants in the flue gas, such as SO_2, to drop out, reducing the need for substantial upgrading of upstream control devices. [18] Using a slurry of ammonium carbonate and ammonium bicarbonate, the solvent absorbs more than 90% of the CO_2 in the flue gas. The resulting CO_2-rich ammonia is regenerated and the CO_2 is stripped from the ammonia mixture under pressure (300 pounds per square inch [psi] compared with 15 psi using MEA), reducing the energy necessary to compress the CO_2 for transportation (generally around 1,500 psi).[19]

The chilled ammonia process is a proprietary process, owned by Alstom. In collaboration with American Electric Power (AEP) and RWE AG (the largest electricity producer in Germany), Alstom began operating its technology on a 20 MW slipstream at AEP's Mountaineer plant in West Virginia in September 2009 and injecting the captured CO_2 into deep saline formations onsite in October 2009.[20] In March 2010, AEP received an award from DOE to cover as much as half the estimated $668 million cost required to scale up the Mountaineer project to capture and store CO_2 from a 235 MW slipstream (approximately 1.5 million tons annually). Once commercial viability is demonstrated at Mountaineer, AEP plans to install the technology at its 450 MW Northeastern Station in Oologah, OK, early in the next decade. The captured gas would be used for EOR. The target is for full commercialization in 2015.

Ammonia (Powerspan)

A second ammonia-based, regenerative process for CO_2 capture from existing coal-fired facilities does not involve chilling the flue gas before it enters the absorber. Using higher flue gas temperatures increases the CO_2 absorption rate in the absorber and, therefore, the CO_2 removal. However, the higher flue gas temperatures also mean that upgrades to existing FGD devices would be necessary.[21]

This process, called ECO_2, is being developed by Powerspan.[22] Powerspan announced test results for its 1 MW pilot unit, located at FirstEnergy Corp's R.E. Burger Plant, in December 2009. The pilot plant tests suggest that carbon capture and compression costs with ECO_2 are below $50 a ton at 90% reduction.[23]

Pre-Combustion CO_2 Capture

Currently, a requirement for the pre-combustion capture of CO_2 is the use of Integrated Gasification Combined-Cycle (IGCC) technology to generate electricity.[24] There are currentlyfour commercial IGCC plants worldwide (two in the United States) each with a capacity of about 250 MW. The technology has yet to make a major breakthrough in the U.S. market because its potential superior environmental performance is currently not required under the Clean Air Act, and as discussed above for carbon capture technology, its higher costs can not be justified (see the Virginia State Corporation Commission decision, discussed below).

Carbon capture in an IGCC facility would happen before combustion, under pressure using a physical solvent (e.g., Selexol and Rectisol processes), or a chemical solvent (e.g., methyl diethanolaimine (MDEA)). A simplified illustration of this process is provided in Figure 2. Basically, the IGCC unit pumps oxygen and a coal slurry into a gasifier to create a syngas consisting of carbon monoxide and hydrogen. The syngas is cleaned of conventional pollutants (SO2, particulates) and sent to a shift reactor which uses steam and a catalyst to produce CO_2 and hydrogen. Because the gases are under substantial pressure with a high CO_2 content, a physical solvent can separate out the CO_2. The advantage of a physical solvent is that the CO_2 can be freed and the solvent regenerated by reducing the pressure—a process that is substantially less energy intensive than having to heat the gas as in an MEA stripper.

From the capture process, the CO_2 is further compressed for transportation or storage, and the hydrogen is directed through gas and steam cycles to produce electricity. MIT estimates the efficiency loss from incorporating capture technology on an IGCC facility is about 19% (from 38.4% efficiency to 31.2%).[25] This loss of efficiency comes in addition to the necessary capital and operations and maintenance cost of the equipment and reagents. For new construction, the estimated increase in electricity generating cost on a levelized basis generally ranges from 22%- 25%, with American Electric Power estimating the cost increase at 41 %.[26]

There is a lot of activity surrounding the further commercialization of IGCC technology and in the demonstration of carbon capture methods on that technology. As illustrated in Figure 3, numerous projects are currently in the development pipeline. The FutureGen initiative—delayed by DOE's decision to restructure the program in early 2008 and subsequently revived in 2009 with funding from P.L. 111-5—may influence how other IGCC projects develop (see box below).[27]

Combustion CO_2 Capture

Attempts to address CO_2 during the combustion stage of generation focus on increasing the CO_2 concentration of the flue gas exiting the boiler. The more concentrated the CO_2 is when it exits the boiler, the less energy (and cost) is required later to prepare it for transport or storage. The most developed approach involves combusting the coal with nearly pure oxygen (>95%) instead of air, resulting in a flue gas consisting mainly of highly concentrated CO_2 and water vapor. Using existing technology, the oxygen would be provided by an air-separation unit—an energy- intensive process that would be the primary source of reduced efficiency. The details of this "oxyfuel" process are still being refined, but generally, from the boiler the exhaust gas is cleaned of conventional pollutants (SO_2, NOx, and particulates) and some of the gases recycled to the boiler to control the higher temperature resulting from coal combustion with pure oxygen. The rest of the gas stream is sent for further purification and compression in preparation for transport and/or storage.[28] Depending on site-specific conditions, oxy-fuel could be retrofitted onto existing boilers. A simplified illustration of this process is provided in **Figure 4**.

The largest oxy-fuel demonstration projects under development are the Vattenfall Project in Germany and the Callide Oxyfuel Project in Queensland, Australia. The Vattenfall project is a 30 MW pilot plant being constructed at Schwarze Pumpe, which began operation in September 2008. The captured CO_2 will be put in geological storage once siting and permitting processes are completed.[29] The Callide Project is being sponsored by CS Energy, who, with six partners, is retrofitting a 30 MW boiler at its Callide-A power station with an oxy-fuel process. Operation of the oxy-fuel process is scheduled for 2011, with transport and geological storage of the CO_2 planned for 2011.[30]

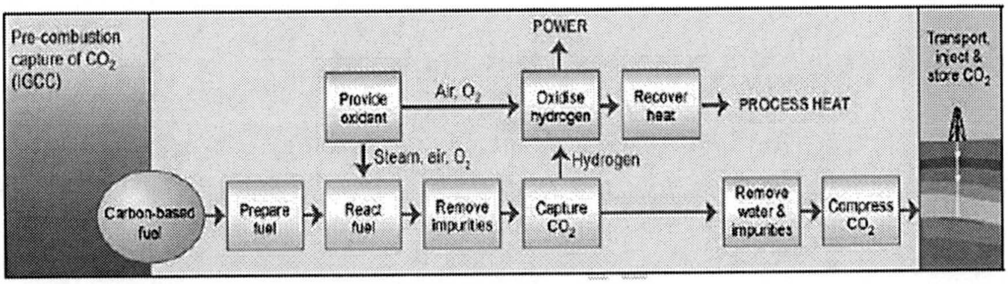

Source: Scottish Centre for Carbon Storage. Figure available at http://www.geos.ed.ac.uk/sccs/capture/ precombustion.html.

Figure 2. Simplified Illustration of Pre-Combustion CO_2 Capture

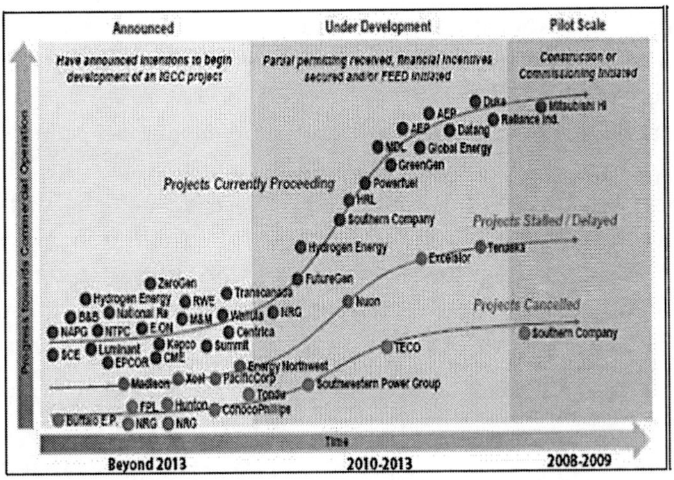

Source: Emerging Energy Research (EER), “Global IGCC Power Markets and Strategies: 2007-2030” (December 2007). See http://www.emerging-energy

Figure 3. Status of Global IGCC Projects

Numerous other bench- and pilot-plant scale initiatives are underway with specific work being conducted on improving the efficiency of the air-separation process. MIT estimates the efficiency losses from the installation of oxy-fuel at 23% for new construction and 3 1%-40% for retrofit on an existing plant (depending on boiler technology).[31] This loss of efficiency comes in addition to the necessary capital and operations and maintenance cost of the equipment and reagents. For new construction, the increase in electricity generating cost on a levelized basis would be about 46%. In the case of retrofit plants where the capital costs are fully amortized, the oxy-fuel capture process would increase generating costs on a levelized basis by about 170%-206%.[32]

DOE-Supported Technology Development

As summarized in Table 2, CO_2 capture technology is currently estimated to significantly increase the costs of electric generation from coal-fired power plants. Research is ongoing to improve the economics and operation of carbon capture technology. DOE's National Energy Technology Laboratory (NETL) is supporting a variety of carbon capture technology research and development (R&D) projects for pre-combustion, oxy-combustion, and post-combustion applications. A detailed description of all the NETL projects, and of carbon capture technology R&D efforts in the private sector, is beyond the scope of this report. However, funding from DOE (described later) is supporting approximately two dozen carbon capture research projects that range from bench-scale to pilot-scale testing. [33] The types of research explored in the NETL-supported projects include the use of membranes, physical solvents, oxy-combustion, chemical sorbents, and combinations of chemical and physical solvents. According to the NETL, these technologies will be ready for slipstream tests by 2014 and for large-scale field testing by 2018.[34]As discussed above, some projects pursued by the private sector are ready for pilot-scale testing (e.g., the AEP Mountaineer project in West Virginia).

ROLES FOR GOVERNMENT

Generally, studies that indicate that emerging, less carbon-intensive new technologies are both available and cost-effective incorporate a price mechanism (such as a carbon tax) that provides the necessary long-term price signal to direct research, development, demonstration, and deployment efforts (called "demand-pull" or "market-pull" mechanisms).[35] Developing such a price signal involves variables such as the magnitude and nature of the market signal, and its timing, direction, and duration. In addition, studies indicate combining a sustained price signal with public support for research and development efforts is the most effective long-term strategy for encouraging development of new technology (called "technology-push" mechanisms).[36] As stated by Richard D. Morgenstern: "The key to a long term research and development strategy is both a rising carbon price, and some form of government supported research program to compensate for market imperfections."[37]

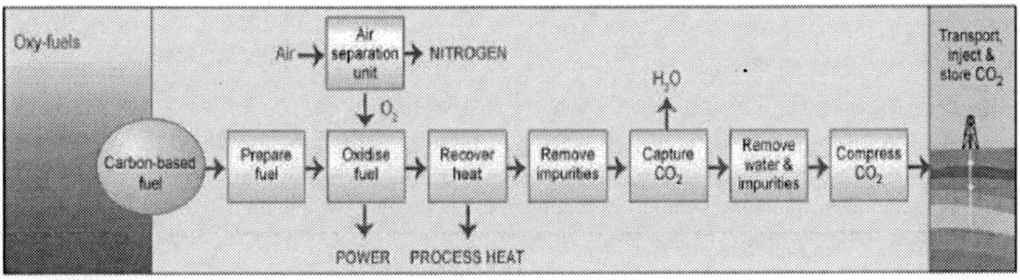

Source: Scottish Centre for Carbon Storage. Figure available at http://www.geos.ed.ac.uk/sccs/capture/ oxyfuel.html.

Figure 4. Simplified Illustration of Oxy-fuels CO_2 Capture

Table 2. MIT Estimates of Additional Costs of Selected Carbon Capture Technology (Percent Increase in Electric Generating Costs on Levelized Basis)

	New Construction	Retrofit[a]
Post-combustion (MEA)	60%-70%	220%-250%
Pre-combustion (IGCC)	22%-25%	not applicable
Combustion (Oxy-fuel)	46%	170%-206%

Source: Massachusetts Institute of Technology, *The Future of Coal: An Interdisciplinary MIT Study* (2007), pp.27, 30, 36, 149. See text for discussion of technologies.

a. Assumes capital costs have been fully amortized.

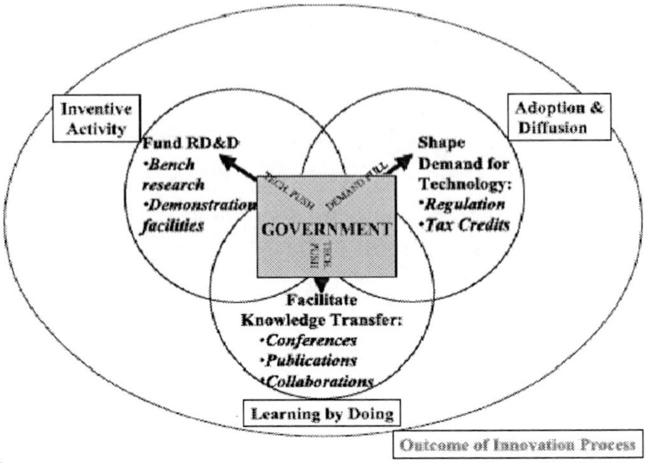

Source: Margaret R. Taylor, Edward S. Rubin and David A Hounshell, "Control of SO2 Emissions from Power Plants: A Case of Induced Technological Innovation in the U.S.," *Technological Forecasting & Social Change* (July 2005), p. 699.

Figure 5. The Federal Role in R&D

The various roles the government could take in encouraging development of environmental technologies are illustrated in Figure 5. The federal role in the innovation process is a complex one, reflecting the complexity of the innovation process itself. The inventive activity reflected by government and private research and development efforts overlap with demand pull mechanisms to promote or require adoption of technology, shaping the efforts. Likewise, these initiatives are facilitated by the government as a forum for feedback gained through both developed and demonstration efforts and practical application. The process is interlinked, overlapping, and dynamic, rather than linear. Attempting to implement one role in a vacuum can result in misdirected funding or mis-timing of results.

This section discusses these different roles with respect to encouraging development of carbon capture technology, including (1) the need for a demand-pull mechanism and possible options; (2)current technology-push efforts at the U.S. Department of Energy (DOE) and the questionstheyraise; and (3) comparison of current energy research and development efforts with past mission-oriented efforts.

THE NEED FOR A DEMAND-PULL MECHANISM

> Economists note that the driving force behind the development of new and improved technologies is the profit motive.... However, market forces will provide insufficient incentives to develop climate-friendly technologies if the market prices of energy inputs do not fully reflect their social cost (inclusive of environmental consequences).... Even if energy prices reflect all production costs, without an explicit greenhouse gas policy firms have no incentive to reduce their greenhouse gas emissions per se beyond the motivation to economize on energy costs. For example, a utility would happily find a way to generate the same amount of electricity with less fuel, but without a policy that makes carbon dioxide emissions costly, it would not care specifically about the carbon content of its fuel mix in choosing between, say, coal and natural gas. For firms to have the desire to innovate cheaper and better ways to reduce emissions (and not merely inputs), they must bear additional financial costs for emissions.[38]

Much of the focus of debate on developing carbon capture technology has been on research, development, and demonstration (RD&D) needs. However, for technology to be fully commercialized, it must meet a market demand—a demand created either through a price mechanism or a regulatory requirement. As suggested by the previous discussion, any carbon capture technology for coal-fired power plants will increase the cost of electricity generation from affected plants with no increase in efficiency. Therefore, widespread commercialization of such technology is unlikely until it is required, either by regulation or by a carbon price. Indeed, regulated industries may find their regulators reluctant to accept the risks and cost of installing technology that is not required by legislation. This sentiment was reflected in a recent decision by the Virginia State Corporation Commission in denying an application by Appalachian Power Company (APCo) for a rate adjustment to construct an IGCC facility:

> The Company asserted that the value of this project is directly related to (1) potential future legal requirements for carbon capture and sequestration; and (2) the proposed IGCC Plant's potential ability to comply cost effectively with any such requirements. Both of these factors, however, are unknown at this time and do not overcome the other infirmities in the Application. The legal necessity of, and the capability of, cost-effective carbon capture and sequestration in this particular IGCC Plant, at this time, has not been sufficiently established to render APCo's Application reasonable or prudent under the Virginia Statute we must follow.[39]

At the same time, there is reluctance to invest in technology that is not required, and the unresolved nature of greenhouse gas regulation is affecting investment in any coal-fired generation. [40] The risk involved in investing in coal-fired generation absent anticipated greenhouse gas regulations is outlined in "The Carbon Principles" announced by three Wall Street banks— Citi, JP Morgan Chase, and Morgan Stanley—in February 2008. As stated in their paper:

> The absence of comprehensive federal action on climate change creates unknown financial risks for those building and financing new fossil fuel generation resources. The Financial Institutions that have signed the Principles recognize that federal CO_2 control legislation is being considered and is likely to be adopted during the service life of many new power plants. It is prudent to take concrete actions today that help developers, investors and financiers to identify, analyze, reduce and mitigate climate risks.[41]

Similarly, lack of a regulatory scheme presents numerous risks to any RD&D effort designed to develop carbon capture technology. Unlike a mission-oriented research effort, like the Manhattan Project to develop an atomic bomb, where the ultimate goal is victory and the cost virtually irrelevant, research efforts focused on developing a commercial device need to know what the market wants in a product and how much the product is worth. At the current time, the market value of a carbon capture device is zero in much of the country because there is no market forcarbon emissions or regulations requiring their reduction.[42]All estimates of value are hypothetical—dependent on a reduction program or regulatory regime that doesn't exist. With no market or regulatory signals determining appropriate performance standards and cost- effectiveness criteria, investment in carbon capture technology is a risky business that could easily result in the development of a "white elephant" or "gold-plated" technology that doesn't meet market demand.

While the "threat" of a carbon regime is stimulating RD&D efforts and influencing decisions about future energy (particularly electricity) supply, the current spread of greenhouse gas control regimes being proposed doesn't provide much guidance in suggesting appropriate performance and cost-effectiveness benchmarks for a solution with respect to coal-fired generation. For example, isolating just one variable in the future price of carbon under a cap-and-trade program— tonnage reduction requirement—the future value of carbon reductions can vary substantially.[43] As illustrated by **Figure 6**, three possible reduction targets in 2050—maintaining current 2008 levels (287 billion metric tons [bmt]), reducing emissions to 50% of 1990 levels (203 bmt), and reducing emissions to 20% of 1990 levels (167 bmt)—result in substantially different price tracks for CO_2.[44] Without a firm idea of the tonnage goal and reduction schedule, any deployment or commercialization strategy would be a high-risk venture, as suggested by the previously noted Virginia State Corporation Commission conclusion.

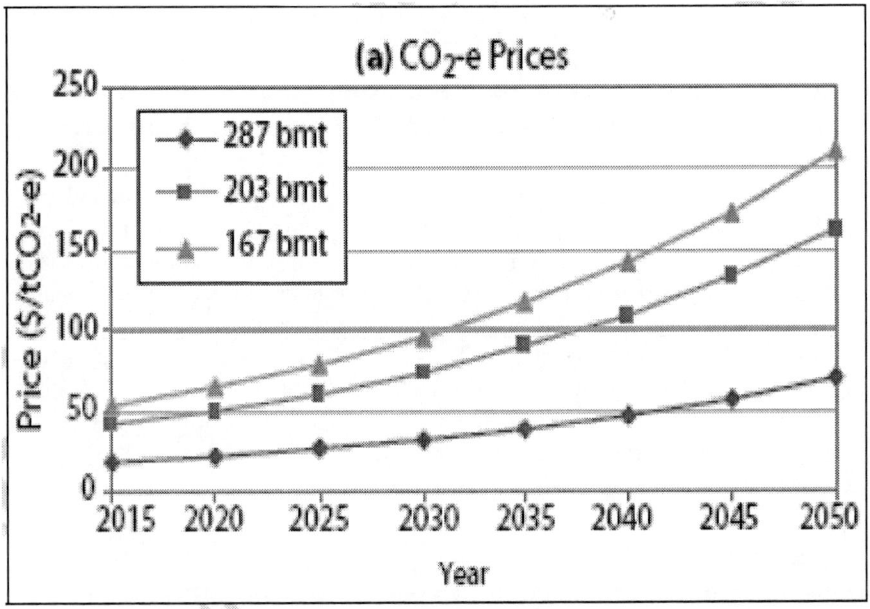

Source: Segey Paltsev, et al., *Assessment of U.S. Cap-and-Trade Proposals*, MIT Joint Program on the Science and Policy of Global Change, Report 146 (April 2007), p. 16. For details on the analysis presented here, consult the report. Available at http://mit.edu/globalchange.

Note: CO_2e = carbon dioxide equivalent

Figure 6. CO_2 Price Projections

APPROACHES TO A DEMAND-PULL MECHANISM

There are two basic approaches to a demand-pull mechanism: (1) a regulatory requirement, and (2) a price signal via a market-based CO_2 reduction program. These approaches are not mutually exclusive and can serve different goals. For example, a regulation focused on new construction (such as the New Source Performance Standard under Section 111 of the Clean Air Act[45]) could be used to phase in deployment of carbon capture technology and prevent more coal-fired facilities from being constructed without carbon capture (or ensure they would be at least "ready" for carbon capture later). At the same time, a carbon tax or cap-and-trade program could be initiated to begin sending a market signal to companies that further controls will be necessary in the future if they decide to continue operating coal-fired facilities.

Creating Demand through a Regulatory Requirement: An Example from the SO_2 New Source Performance Standards

> It is an understatement to say that the new source performance standards promulgated by the EPA were technology-forcing. Electric utilities went from having no scrubbers on their generating units to incorporating very complex chemical processes. Chemical plants and refineries had scrubbing systems that were a few feet in diameter, but not the 30- to 40-foot diameters required by the utility industry. Utilities had dealt with hot flue gases, but not with saturated flue gases that contained all sorts of contaminants. Industry, and the US EPA, has always looked upon new source performance standards as technology-forcing, because they force the development of new technologies in order to satisfy emissions requirements.[46]

The most direct method to encourage adoption of carbon capture technology would be to mandate it. Mandating a performance standard on coal-fired power plants is not a new idea; indeed, Section 111 of the Clean Air Act requires the Environmental Protection Agency (EPA) to develop New Source Performance Standards (NSPS) for any new and modified power plant (and other stationary sources) that in the Administrator's judgment "causes, or contributes significantly to, air pollution which may reasonably be anticipated to endanger public heath or welfare." NSPS can be issued for pollutants for which there is no National Ambient Air Quality Standard (NAAQS), like carbon dioxide.[47] In addition, NSPS is the floor for other stationary source standards such as Best Available Control Technology (BACT) determinations for Prevention of Significant Deterioration (P SD) areas and Lowest Achievable Emission Rate (LAER) determinations for non-attainment areas.[48]

The process of forcing the development of emission controls on coal-fired power plants is illustrated by the 1971 and 1978 SO_2 NSPS for coal-fired electric generating plants. The Clean Air Act states that NSPS should reflect "the degree of emission limitation achievable through theapplication of the best system of emission reduction which (taking into account the cost of achieving such reductions and any non-air quality health and environmental impact and energy requirements) the Administrator determines has been adequately demonstrated."[49] In promulgating its first utility SO_2 NSPS in 1971, EPA determined that a 1.2 pound of SO_2 per million Btu of heat input performance standard met the criteria of Sec. 111—a standard that required, on average, a 70% reduction in new power plant emissions, and could be met by low- sulfur coal that was available in both the eastern and western parts of the United States, or by the use of emerging flue gas desulfurization (FGD) devices.[50]

At the time the 1971 Utility SO2 NSPS was promulgated, there was only one FGD vendor (Combustion Engineering) and only three commercial FGD units in operation—one of which would be retired by the end of the year.[51] This number would increase rapidly, not only because of the NSPS, but also because of the promulgation of the SO_2 NAAQS, the 1973 Supreme Court decision preventing significant deterioration of pristine areas,[52] and state requirements for stringent SO_2 controls, which opened up a market for retrofits of existing coal-fired facilities in addition to the NSPS focus on new facilities. Indeed, most of the growth in FGD installations during the early and mid-1970s was in retrofits—Taylor estimates that between 1973 and 1976, 72% of the FGD market was in retrofits.[53] By 1977, there were 14 vendors offering full-scale commercial FGD installation.[54]

Despite this growth, only 10% of the new coal-fired facilities constructed between 1973 and 1976 had FGD installations. In addition, the early performance of these devices was not brilliant.[55] In 1974, American Electric Power (AEP) spearheaded an ad campaign to have EPA reject FGD devices as "too unreliable, too impractical for electric utility use" in favor of tall stacks, supplementary controls, and low-sulfur western coal.[56] This effort was ultimately unsuccessful as the Congress chose to modify the NSPS requirements for coal-fired electric generators in 1977 by adding a "percentage reduction" requirement. As promulgated in 1979, the revised SO_2 NSPS retained the 1971 performance standard but added a requirement for a 70%-90% reduction in emissions, depending on the sulfur content of the coal.[57]At the time, this requirement could be met only through use of an FGD device. The effect of the "scrubber requirement" is clear from the data provided in Figure 7. Based on their analysis of FGD development, Taylor, Rubin, and Hounshell state the importance of demand-pull instruments:

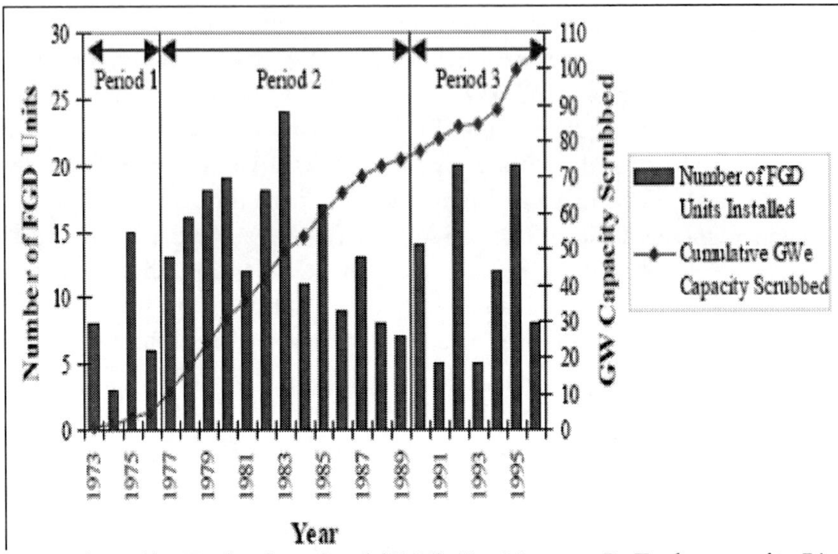

Source: Adapted by Taylor from Soud (1994). See Margaret R. Taylor, op. cit., 74.
Note: Numbers are archival through June 1994, then projected for 1994-96.

Figure 7. Number of FGD Units and Cumulative GW Capacity of FGD Units: 1973-1996

> Results indicate that: regulation and the anticipation of regulation stimulate invention; technology-push instruments appear to be less effective at prompting invention than demand- pull instruments; and regulatory stringency focuses inventive activity along certain technology pathways.[58]

That government policy could force the development of a technology through creating a market should not suggest that the government was limited to that role, or that the process was smooth or seamless. On the latter point, Shattuck, et al., summarize the early years of FGD development as follows:

> The Standards of Performance for New Sources are technology-forcing, and for the utility industry they forced the development of a technology that had never been installed on facilities the size of utility plants. That technology had to be developed, and a number of installations completed in a short period of time. The US EPA continued to force technology through the promulgation of successive regulations. The development of the equipment was not an easy process. What may have appeared to be the simple

application of an equipment item from one industry to another often turned out to be fraught with unforeseen challenges.[59]

The example indicates that technology-forcing regulations can be effective in pulling technology into the market—even when there remains some operational difficulties for that technology. The difference for carbon capture technology is that for long-term widespread development, a new infrastructure of pipelines and storage sites may be necessary in addition to effective carboncapture technology. In the short-term, suitable alternatives, such as enhanced oil recovery needs and in-situ geologic storage, may be available to support early commercialization projects without the need for an integrated transport and storage system. Likewise, with economics more favorable for new facilities than for retrofits, concentrating on using new construction to introduce carbon capture technology might be one path to widespread commercialization. As an entry point to carbon capture deployment, a regulatory approach such as NSPS may represent a first step, as suggested by the SO_2 NSPS example above.

Creating Demand through a Price Signal: Carbon Taxes, Allowance Pricing, and Auctions

Much of the current discussion of developing a market-pull mechanism for new carbon capture technology has focused on creating a price for carbon emissions. The literature suggests that this is an important component for developing new technology, perhaps more important even than research and development. As stated by the Congressional Budget Office (CBO):

> Analyses that consider the costs and benefits of both carbon pricing and R&D all come to the same qualitative conclusion: near-term pricing of carbon emissions is an element of a cost- effective policy approach. That result holds even though studies make different assumptions about the availability of alternative energy technologies, the amount of crowding out caused by federal subsidies, and the form of the policy target (maximizing net benefits versus minimizing the cost of reaching a target).[60]

Two basic approaches can be employed in the case of a market-based greenhouse gas control program: a carbon tax and a cap-and-trade program. The carbon tax would create a long-term price signal to stimulate innovation and development of new technology. This price signal could be strengthened if the carbon tax were escalated over the long run—either by a statutorily determined percentage or by an index (such as the producer price index). A carbon tax's basic approach to controlling greenhouse gas emissions is to supply the marketplace with a stable, consistent price signal—a signal that would also inform innovators as to the cost performance targets they should seek in developing alternative technologies. Designed appropriately, there would be little danger of the price spikes or market volatility that can occur in the early stages of a tradeable permit program.[61]

A cap-and-trade program creates a price signal for new technology through a market price for carbon permits (called allowances)—an allowance is a limited authorization to emit one metric ton of carbon dioxide equivalent (CO_2e). In a cap-and-trade system, these allowances are issued by the government and either allocated or auctioned to affected companies who may use them to comply with the cap, sell them to other companies on the market, or bank them for future use or sale. The resulting market transactions result in an allowance price. This price on carbon emissions, however, can be both uncertain and volatile.[62] In addition, a low allowance price may be insufficient to encourage technology development and refinement. For example, the 1990 acidrain control program effectively ended the development of FGD for retrofit purposes by setting an emission cap that resulted in low allowance prices and that could be met through the use of low- sulfur coal. Noting that only 10% of phase 1 facilities chose FGD to comply with itsrequirements, Taylor, et al., state:

> The 1990 CAAA, however, although initially predicted to increase demand for FGD systems, eroded the market potential for both dry and wet FGD system applications at existing power plants when the SO_2 allowance trading market returned low-sulfur coal to its importance in SO_2 control.... As a result, research in dry FGD technology declined significantly. In this case, the flexibility provided by the 1990 acid rain regulations discouraged inventive activity in technologies that might have had broader markets under the traditional command-and-control regimes in place prior to 1990.[63] [footnotes from original text omitted]

A cap-and-trade program need not have such a result. For example, to more effectively promote carbon capture technology, the price signal under a greenhouse gas reduction program could be strengthened by requiring the periodic auctioning of a substantial portion of available allowances rather than giving them away at no cost. The SO_2 program allocated virtually all of its allowance at no cost to affected companies. Auctioning a substantial portion of available allowances could create a powerful price signal and provide incentives for deploying new technology if structured properly.[64]The program could create a price floor to facilitate investment in new technology via a reserve price in the allowance auction process. In addition, the stability of that price signal could be strengthened by choosing to auction allowances on a frequent basis, ensuring availability of allowances close to the time of expected demand and making any potential short-squeezing of the secondary market more difficult.[65]

One positive aspect of the acid rain cap-and-trade experience for encouraging deployment of technology was the effectiveness of "bonus" allowances and deadline extensions as incentives to install FGD. Specifically, about 3.5 million of the allowances were earmarked for Phase 1 power plants choosing to install 90% control technology (such as FGD). Such units were allowed to delay Phase 1 compliance from 1995 to 1997 and receive two allowances for each ton of SO_2 reduced below a 1.2 lb. per mmBtu level during 1997-1999. The 3.5 million allowance reserve was fully subscribed, and was a major factor in what FGD was installed during Phase 1 of the program. This experience may bode well for proposed CCS "bonus allowance" provisions in several greenhouse gas reduction schemes currently introduced in the Congress.[66]

CURRENT TECHNOLOGY-PUSH MECHANISMS: DOEINVESTMENT IN CCS R&D

The Department of Energy (DOE) is engaged in a variety of activities to push development and demonstration of carbon capture technologies. These activities include direct spending on research and development, and providing loan guarantees and tax credits to promote carbon capture projects. These technology-push incentives, and the issues they raise, are discussed below.

Direct Spending on R&D

The federal government has recognized the potential need for carbon capture technology—as part of broader efforts to address greenhouse-gas induced climate change—since at least 1997, when DOE spent approximately

$1 million for the entire CCS program.[67] Table 3 shows that DOE programs that provide funding for CCS-related activities totaled nearly $575 million for FY2009, a significant increase since 1997.[68] Funding for CCS R&D increased by nearly 58% from FY2008 to FY2009. Table 3 also shows that funding appears to have decreased for CCS R&D in FY2010 as compared to FY2009, and the request for FY2011 is slightly lower than the FY2010 amounts. However, funding from P.L. 111-5, the American Recovery and Reinvestment Act of 2009 (ARRA, shown in the last column of the table), increases *overall* spending for CCS dramatically compared to previous years.

Table 3. Funding for CCS-Related Activities at DOE($ Thousands)

	FY2008	FY2009	FY2010	FY2011 request	ARRA
Clean Coal Power Initiative (CCPI)[a]	67,444	288,174	0	0	800,000
FutureGen[b]	72,262	0	0	0	1,000,000
Innovation for Existing Plants (IEP)[c]	35,083	48,600	52,000	65,000	—
Advanced Integrated Gasification Combined Cycle[d]	52,029	63,409	63,000	55,000	—
Advanced Turbines[e]	23,125	27,216	32,000	31,000	—
Industrial Carbon Capture Projects	—	—			1,520,000
Site Characterization, Training, Program Direction	—	—			80,000
Subtotal	249,943	427,399	147,000	151,000	3,400,000
Carbon Sequestration Greenhouse Gas Control	105,985	132,192	140,000	133,000	—
Carbon Sequestration Energy Innovation Hub	0	0	0		—
Carbon Sequestration Focus Area for	9,635	13,608	14,000	10,000	—
Carbon Sequestration Science					
Subtotal for Carbon Sequestration	115,620	145,800	154,000	143,000	—
Total	365,563	573,199	301,000	294,000	3,400,000

Source: CRS, from the U.S. Department of Energy, FY2011 *Congressional Budget Request*, volume 3, Fossil EnergyResearch and Development, at http://www.cfo.doe.gov/budget/11budget/Content/Volume%203.pdf; and U.S.

Department of Energy, FY2010 *Congressional Budget Request*, volume 7, Fossil Energy Research and Development, at http://www.cfo.doe.gov/budget/10budget/Content/Volumes/Volume7.pdf.

Notes: Overall FY20 10 Fossil Energy Research appropriations are included in CRS Report R40669, *Energy and Water Development: FY2010 Appropriations*.

a. The FY2011 budget request does not include any funds for CCPI demonstration projects because $800 million is already provided by ARRA (P.L. 111-5) for Phase III of the CCPI program.

b. Language in ARRA indicated that $1 billion would be allocated for Fossil Energy R&D. On June 12, 2009, Secretary Chu announced that the funds would be used to support FutureGen.

c. In its FY2011 budget request, DOE indicates that all the IEP activity is focused on the development of post- combustion CO_2 capture technology for new and existing plants.

d. According to DOE, the IGCC activity is focused on developing advanced gasification-based technologies to reduce the costs of near-zero emissions (including CO_2) coal-based IGCC plants. The program is also intended to improve the thermal efficiency of the plants, and to achieve near-zero atmospheric emissions for all pollutants, including CO_2, SO_2, NOx, and mercury.

e. The Advanced Turbines program is focused on creating the technology base for turbines that will permit the design of near-zero atmospheric emission IGCC plants (including CO_2).

DOE indicated in its FY2011 budget request that its Office of Fossil Energy will propose a new budget structure for FY2012 to reflect the increased focus on CCS technologies within the DOE Clean Coal program. The FY2011 budget request further stated that research, development, and deployment of CCS is a major component of global activities needed if coal power plants with CCS are to be deployed "in a timeframe consistent with climate stabilization goals."[69] DOE indicated that these activities may lead to mass commercial deployment of CCS by 2020.[70]

Carbon Capture and Sequestration in the American Recovery and Reinvestment Act of 2009 (ARRA)

Funding for carbon capture and sequestration technology has increased substantially as a result of enactment of ARRA (P.L. 111-5). In the compromise legislation considered in conference on February 11, 2009, the conferees agreed to provide $3.4 billion through FY20 10 for fossil energy research and development within the Department of Energy (DOE). Of that amount, $1.52 billion would be made available for a competitive solicitation for industrial carbon capture and energy efficiency improvement projects, according to the explanatory statement accompanying the legislation. This provision likely refers to a program for large scale demonstration projects that capture CO_2 from a range of industrial sources. A small portion of the $1.52 billion would be allocated for developing innovative concepts for reusing CO_2, according to the explanatory statement. Of the remaining $1.88 billion, $1 billion would be available for fossil energy research and development programs. The explanatory statement did not specify which program or programs would receive funding, however, or how the $1 billion would be allocated. However, onJune 12, 2009, Energy Secretary Chu announced that the $1 billion would be used to support a renewed FutureGen facility in Mattoon, IL. Of the remaining $880 million, the conferees agreed to allocate $800 million to the DOE Clean Coal Power Initiative Round III solicitations, which specifically target coal-based systems that capture and sequester, or reuse, CO_2 emissions. Lastly, $50 million would be allocated for site characterization activities in geologic formations (for the storage component of CCS activities), $20 million for geologic sequestration training and research, and $10 million for unspecified program activities.

With the announcement that $1 billion of the ARRA funds would be used to restart FutureGen, nearly all of the $3.4 billion agreed to by conferees will be used for CCS activities, which represents a substantial infusion of funding compared to current spending levels. This also amounts to a large and rapid increase in funding over what DOE has spent on CCS *cumulatively* since FY1997.[71] Moreover, the bulk of DOE's CCS program will shift to the capture component of CCS, unless funding for the storage component increases commensurately in annual appropriations. The large and rapid increase in funding, compared to the magnitude and pace of previous CCS spending, may raise questions about the efficacy of a "crash" CCS program as part of a long-term strategy to reduce CO_2 emissions. This issue is discussed further below.

Loan Guarantees and Tax Credits

Appropriations represent one mechanism for funding carbon capture technology R&D and deployment; others include loan guarantees and tax credits, both of which are available under current law.

Loan Guarantees

Loan guarantee incentives that could be applied to CCS were authorized under Title XVII of the Energy Policy Act of 2005 (EPAct2005, P.L. 109-58, 42 U.S.C. §§16511-16514), and were given indefinite authorization under the Omnibus Appropriations Act, 2009 (P.L. 111-8). Title XVII of EPAct2005 authorizes the Secretary of Energy to make loan guarantees for projects that, among other purposes, avoid, reduce, or sequester air pollutants or anthropogenic emissions of greenhouse gases. The Omnibus Appropriations Act for FY2009 restates the loan guarantee authority and provides $6 billion in loan guarantees for coal-based power generation and industrial gasification activities at retrofitted and new facilities that incorporate CCS or other beneficial uses of carbon. The act provides an additional $2 billion in loan guarantees for advanced coal gasification.[72]

Tax Credits

Title XIII of EPAct2005 provided for tax credits that could be used for Integrated Gasification Combined Cycle (IGCC) projects and for projects that use other advanced coal-based generation technologies (ACBGT). For these types of projects, the aggregate credits available under EPAct2005 totaled up to $1.3 billion: $800 million for IGCC projects, and $500 million forACBGT projects. Qualifying projects under Title XIII of EPAct2005 were not limited to technologies that employ carbon capture technologies, but the Secretary of the Treasury was directed to give high priority to projects that include greenhouse gas capture capability. An additional $350 million of tax credits were made available for coal gasification projects.

Sections 111 and 112 of P.L. 110-343, Division B, the Energy Improvement and Extension Act of 2008 (part of the Emergency Economic Stabilization Act of 2008), increased the aggregate tax credits available from $1.65 billion to $3.15 billion. Section 111 added an additional $1.25 billion to the existing tax credit authority for ACBGT projects. Section 112 added an additional $250 million to $350 million in existing authority for the coal gasification investment credit, for gasification projects that separate and sequester at least 75% of the project's total CO_2 emissions.

Section 115 of the act added a new tax credit for sequestering CO_2 and storing it underground. The section provides for a credit of $20 per metric ton of CO_2 captured at a qualified facility and disposed of in secure geological storage, and $10 per metric ton if the CO_2 is used as a tertiary injectant for the purposes of enhanced oil or natural gas recovery. To qualify for the tax credit, the facility must capture at least 500,000 metric tons of CO_2 per year. If CO_2 is used for enhanced oil or gas recovery, a tax credit would be available only for an initial injection; CO_2 subsequently recaptured, recycled, and re-injected would not be eligible for a tax credit.

ENCOURAGING TECHNOLOGY DEVELOPMENT IN THEABSENCE OF A MARKET: ISSUES FOR CURRENT CARBONCAPTURE RD&D POLICY

Each of the funding mechanisms described above—appropriations, loan guarantees, and tax credits—are examples of government "pushing" carbon capture technologies (the upper left arrow in **Figure 5**) via direct spending and through private sector incentives. Thus far, however, these activities are taking place in a vacuum with respect to a carbon market or a regulatory structure. Lacking a price signal or regulatory mandate, it is difficult to assess whether a government-push approach is sufficient for long-term technology development.[73] Some studies appear to discount the necessity of a price signal or regulatory mandate, at least initially, and place a higher priority on the successful demonstration of large-scale technological, economic, and environmental performance of technologies that comprise all of the components of an integrated CCS system: capture, transportation, and storage.[74] So far, however, the only federally sponsored, fully integrated, large-scale CCS demonstration project—called FutureGen (see box below)—failed in its original conception, which may have been due, in part, to the lack of a perceived market.

DOE announced it was restructuring the FutureGen program because of its rising costs, which are difficult to assess against the project's "benefits" without a monetary value attached to those benefits (i.e., the value of carbon extracted from the fuel and permanently sequestered). A carbon market would at least provide some way of comparing costs against benefits. One could arguethat the benefits of CCS accrue to the amelioration of future costs of environmental degradation caused by greenhouse gas-induced global warming. Although it may be possible to identify overall environmental benefits to removing CO_2 that would otherwise be released to the atmosphere, assigning a monetary value to those benefits to compare against costs is extremely difficult.

TRYING TO PICK A WINNER: FUTUREGEN

On February 27, 2003, President Bush proposed a 10-year, $1 billion project to build a coal-fired power plant that integrates carbon sequestration and hydrogen production while generating 275 megawatts of electricity, enough to power about 150,000 average U.S. homes. As originally conceived, the plant would have been a coal-gasification facility and would have produced between 1 and 2 million metric tons of CO_2 annually. The plant was envisioned to be nearly emissions-free because most of the CO_2 produced would be captured and sequestered underground. On January 30, 2008, DOE announced that it was "restructuring" the FutureGen program away from a single, state-of- the-art "living laboratory" of integrated R&D technologies—a single plant—to instead pursue a new strategy of providing funding for the addition of CCS technology to multiple commercial-scale Integrated Gasification Combined Cycle (IGCC) power plants. In the restructured program, DOE would support up to two or three demonstration projects, each of at least 300 MW, and that would sequester at least 1 million metric tons of CO_2 per year. In its budget justification for FY2009, DOE cited "new market realities" for its decision, namely rising material and labor costs for new power plants and the need to demonstrate commercial viability of IGCC plants with CCS. A policy question that emerged following the DOE's decision to scrap the original FutureGen concept was whether such a concept can be viable without a long-term price signal for carbon. FutureGen supporters have indicated that the rise in FutureGen's projected costs were consistent with the rise in global energy infrastructure projects due to inflation, implying that rising costs are not unique to FutureGen. Nevertheless, the reasons given by DOE in its decision to cancel the original concept are *prima facie* evidence that lack of a price signal for carbon in the face of known and rising costs for plant construction created too much uncertainty for the agency to continue the project. It is unclear whether a long-term price signal would have supported the FutureGen concept anyway, given the project's other uncertainties, such as its choice of a capture technology and disagreements over the private cost-share agreement.

With the passage of ARRA, however, FutureGen was revived. On June 12, 2009, Secretary Chu announced that the $1 billion of funding from ARRA will be used to support FutureGen, and that the plant will be built in Mattoon, IL, the site selected by the FutureGen Alliance in 2007. According to DOE, its total anticipated contribution to FutureGen will be $1.073 billion, and the FutureGen Alliance will contribute between $400 and $600 million to the project. On July 14, 2009, DOE issued a National Environmental Policy Act record of decision to move forward on FutureGen. Some reports indicate that the newly revived plans for FutureGen call for an initial carbon capture goal of 60% for the facility, with the ultimate goal of achieving a 90% capture rate—the target set in the project's original conception. Some environmental groups have expressed views that the lower capture rate may put FutureGen in the same category as other CCS commercialization projects, calling into question the status of FutureGen as a "flagship facility to demonstrate carbon capture and storage at commercial scale."

Sources: DOE, http://www.fossil DOE_Announces_Restructured_FutureG.htm; DOE FY2009 Budget request, p. 16; FutureGen Alliance press release (April 15, 2008), http://www.futuregenalliance.org /news/releases/pr_04-15-08.stm; Ben Geman, "Enviros Fault Scaled- Back FutureGen Carbon Goal," *Greenwire*, June 16, 2009; Secretary Chu's announcement on FutureGen at http://www.energy.gov/news2009/7454.htm; Michael T. Burr, "Death of a Turkey, DOE's Move to 'Restructure' FutureGen Clears the Way for a More Rational R&D," Public Utilities Fortnightly (March 2008); and David Goldston, "Demonstrably Wrong," *Nature*, vol. 453, no. 16 (April 30, 2008), p. 16.

What Should the Federal Government Spend on CCS Technology Development?

As discussed above, several studies underscore the value of a long-term price or regulatory signal to shape technological development and, presumably, to help determine a level of federal investment needed to encourage commercialization of an environmental technology such as carbon capture. As stated by Fischer:

> With respect to R&D for specific applications (such as particular manufacturing technologies or electricity generation), governments are notoriously bad at picking winners... [e.g., the breeder reactor]. The selection of these projects is best left to private markets while the government ensures those markets face the socially correct price signals.[75]

Despite the lack of regulatory incentives or price signals, DOE has invested millions of dollars since 1997 into capture technology R&D, and the question remains whether it has been too much, too little, or about the right amount. In addition to appropriating funds each year for the DOE program, Congress signaled its support for RD&D investment for CCS through provisions for tax credits available for carbon capture technology projects and through loan guarantees. Congress also authorized a significant expansion of CCS spending at DOE in the Energy Independence and Security Act of 2007 (EISA, P.L. 110-140), which would authorize appropriations for a total of $2.2 billion from FY2008 through FY20 13. Although EISA places an increased emphasis on large-scale underground injection and storage experiments, the legislation authorizes $200 million per year for projects that demonstrate technologies for the large-scale capture of CO_2 from arange of industrial sources. Lastly, ARRA provides a dramatic infusion of funding for CCS RD&D over the next several years.

Legislation in the 110th and 111th Congresses

Legislation introduced in the 110th Congress would have authorized specific amounts of spending for CCS and capture technology development. Notably, the Carbon Capture and Storage Early Deployment Act (H.R. 6258) would have authorized distribution utilities[76] to collect an assessment on fossil-fuel based electricity delivered to retail customers. The assessment would total approximately $1 billion annually, and would be issued by a corporation—established by referendum among the distribution utilities—as grants or contracts to private, academic, or government entities to accelerate commercial demonstration or availability of CO_2 capture and storage technologies and methods. This legislation contained elements that resembled, in many respects, recommendations offered in the MIT report.[77] Hearings were held, but the measure was not reported out of committee.

Other bills introduced in the 110th Congress included incentives such as tax credits, debt financing, and regulations to promote CO_2 capture technology development. For example, S. 3132, the Accelerating Carbon Capture and Sequestration Act of 2008, would have provided a tax credit of $20 per metric ton of CO_2 captured and stored.[78] S. 3233, the 21st Century Energy Technology Deployment Act, would have established a corporation that could issue debt instruments (such as bonds) for financing technology development. A priority cited in S. 3233 was the deployment of commercial-scale CO_2 capture and storage technology that could capture 10 million short tons of CO_2 per year by 2015. A bill aimed at increasing the U.S. production of oil and natural gas while minimizing CO_2 emissions, the American Energy Production Act of 2008 (S. 2973), called for the promulgation of regulations for clean, coal-derived fuels. Facilitiesthat process or refine such fuels would be required to capture 100% of the CO_2 that would otherwise be released at the facility. None of the bills were enacted into law.

In the 111th Congress, H.R. 2454, the American Clean Energy and Security Act of 2009, and S. 1733, the Clean Energy Jobs and American Power Act, are the two primary energy and climate change legislative proposals. H.R. 2454 passed the House on June 26, 2009, and S. 1733 was ordered to be reported by the Senate Environment and Public Works Committee on November 5, 2009. The CCS provisions in both bills are similar (some sections are identical), and both bills appear to share the goal of fostering the commercial development and deployment of CCS projects as an important component of mitigating greenhouse gas emissions. Both bills contain similar provisions that would create a program to accelerate the commercial availability of CO_2 capture and storage technologies and methods by awarding grants, contracts, and financial assistance to electric utilities, academic institutions, and other eligible entities (similar to H.R. 6258 of the 110th Congress, discussed above). Both bills would also create a second program that would distribute emission allowances from the cap-and-trade provisions to qualifying electric generating plants and industrial facilities.[79]

One bill introduced in the 111th Congress, the New Manhattan Project for Energy Independence (H.R. 513), calls for a system of grants and prizes for RD&D on the scale of the original Manhattan project, with a goal of attaining energy independence for the nation. Other legislation introduced in the 110th Congress invoked the symbolism of the Apollo program of the 1 960s to frame proposals for large-scale energy policy initiatives that include developing CCS technology.[80] The relevance and utility of large-scale government projects, such as the Apollo program, or the Manhattan project, to developing carbon capture technology are explored in the following sections.

Should the Federal Government Embark on a "Crash" Research and Development Program?

Some policymakers have proposed that the United States invest in energy research, development, and demonstration activities at the same level of commitment as it invested in the past during the Manhattan project and the Apollo program.[81] As analogues to the development of technologies to reduce CO_2 emissions and thwart long-term climate change, the Manhattan project and Apolloprogram are imperfect at best. They both had short-term goals, their success or failure was easily measured, and perhaps most importantly, they did not depend on the successfulcommercialization of technology and its adoption by the private sector. Nevertheless, both projects provide a funding history for comparison against CO_2 capture technology cost projections, and as examples of large government-led projects initiated to achieve a national goal. The Manhattan project and Apollo program are discussed briefly below.

The federal government's efforts to promote energy technology development in response to the energy crisis of the 1970s and early 1 980s may be a richer analogy to CO_2 capture technology development than either the Manhattan project or Apollo program. After the first oil crisis in 1973, and with the second oil crisis in the late 1970s, the national priority was to reduce dependence on foreign supplies of energy, particularly crude oil, through a combination of new domestic supplies (e.g., oil shale), energy efficiency technologies, and alternative energy supplies such as solar, among others. The success of these efforts was to have been determined, in part, by the commercialization of energy technologies and alternative energy supplies and their incorporation into American society over the long term. Similarly, many analysts see the development of CCS technology as a necessary step needed over the next several decades or half- century to help alleviate human-induced climate change, which is itself viewed as a global problem for at least the next century or longer. As discussed more fully later, the outcome of the federal government's efforts to promote energy technologies in the 1970s and 1 980s may be instructive to current approaches to develop CCS technology.

The Manhattan Project and Apollo Program

The Manhattan project took place from 1942 to 1946.[82] In July 1945, a bomb was successfully tested in New Mexico, and used against Japan at two locations in August 1945. In 1946, the civilian Atomic Energy Commission was established to manage the nation's future atomic activities, and the Manhattan project officially ended. According to one estimate, the Manhattan project cost \$2.2 billion from 1942-1946 (\$22 billion in 2008 dollars), greater than the original cost and time estimate of approximately \$148 million for 1942 to 1944.[83]

The Apollo program encompassed 17 missions including six lunar landings that took place from FY1960 to FY1973.[84] Although preliminary discussions regarding the Apollo program began in 1960, Congress did not decide to fund it until 1961 after the Soviets became the first country to send a human into space. The peak cost for the Apollo program occurred in FY1966 when NASA's total budget was \$4.5 billion and its funding for Apollo was \$3.0 billion.[85] According to NASA, the total cost of the Apollo program for

FY1 960-FY1 973 was $19.4 billion ($97.9 billionin 2008 dollars).[86] The first lunar landing took place in July 1969. The last occurred in December 1972. Figure 8 shows the funding history for both the Manhattan project and Apollo program.

DOE-Supported Energy Technology Development

The Department of Energy has its origins in the Manhattan project,[87] and became a Cabinet-level department in 1977,[88] partly in response to the first oil crisis of 1973, caused in part by the Arab oil embargo. Another oil crisis (the "second" oil crisis) took place from 1978-1981 as a result of political revolution in Iran. Funding for DOE energy R&D rose in the 1970s in concert with high oil prices and resultant Carter Administration priorities on conservation and development of alternative energy supplies. Crude oil prices fell during the 1980s and the Reagan Administration eliminated many energy R&D programs that began during the oil crisis years. **Figure 8** shows the rise and fall of funding for DOE energy technology programs from 1974 to 2008.

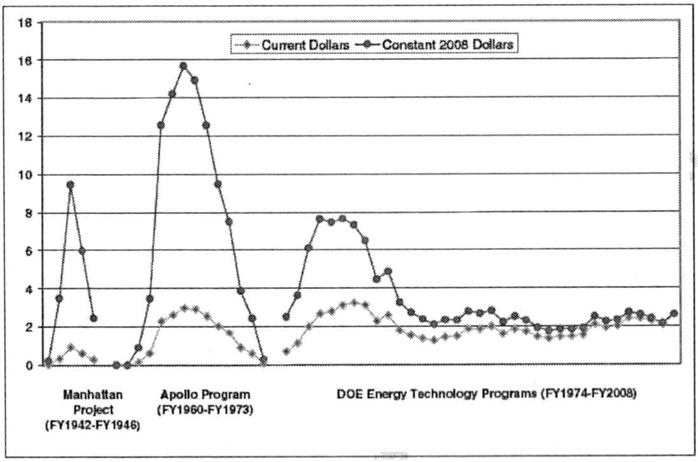

Source: Congressional Research Service. Manhattan Project data: Richard G. Hewlett and Oscar E. Anderson, Jr., A *History of the United States Atomic Energy Commission: The New World, 1939/1946, Volume I.* Apollo program data: Richard Orloff, *Apollo By The Numbers: A Statistical Reference*, NASA SP-2000-4029, 2004 web update. DOE data: CRS Report RS22858, *Renewable Energy R&D Funding History: A Comparison with Funding for Nuclear Energy, Fossil Energy, and Energy Efficiency R&D*, by Fred Sissine.

Figure 8. Annual Funding for the Manhattan Project, Apollo Program, and DOE Energy Technology Programs

Comparisons to CO_2 Capture R&D at DOE

Current DOE spending on CCS technology development (discussed above) is far below levels of funding for the Manhattan project and Apollo program and for the energy technology R&D programs at their peak spending in the late 1970s and early 1980s. The development of CO_2 capture technology is, of course, only one component of all federal spending on global climate change mitigation. However, the total annual federal expenditures on climate change, including basic research, are still far less than the Manhattan project and Apollo program, although similar to DOE energy technology development programs during their peak spending period.[89] For comparison, the FY2008 budget and FY2009 budget request for DOE's energy technology R&D is approximately $3 billion per year. (See **Figure 8**.)

Even if spending on CO_2 capture technology were increased dramatically to Manhattan project or Apollo program levels, it is not clear whether the goal of developing a commercially deployable technology would be realized. As mentioned above, commercialization of technology and integration of technology into the private market were not goals of either the Manhattan project or Apollo program. For the Manhattan project, it did not matter what the cost was, in one sense, if a consequence of failing to build a nuclear weapon was to lose the war. For CO_2 capture, the *primary* goal is to develop a technology that would be widely deployed and thus effective at removing a substantial amount of CO_2 over the next half century or more, which necessarily requires its commercialization and widespread use throughout the utility sector.

The Possibility of Failure: The Synthetic Fuels Corporation

A careful study of one of the federal projects initiated in response to the energy crisis of the 1970s and early 1980s—the Synthetic Fuels Corporation (SFC)—may provide a valuable comparison to current thinking about the federal role in CO_2 capture technology development:

> The government's attempt to develop a synthetic fuels industry in the late 1970s and early 1 980s is a case study of unsuccessful federal involvement in technology development. In 1980, Congress established the Synthetic Fuels Corporation (SFC), a quasi-independent corporation, to develop large-scale projects in coal and shale liquefaction and gasification. Most of the projects centered on basic and conceptual work that would contribute to demonstration programs in later stages, although funds were expended on several prototype and full-scale demonstration experiments.

> Formed in response to the 1970s energy crisis, the SFC was intended to support projects that industry was unable to support because of technical, environmental, or financial uncertainties. Federal loans, loan guarantees, price guarantees, and other financial incentives totaling $20 billion were authorized to spur industry action. Although SFC was designed to continue operating until at least 1992, the collapse in energy prices, environmental concerns, lack of support from the ReaganAdministration, and administrative problems ended the synthetic fuels program in 1986.[90] [citations from original text omitted]

One of the primary reasons commonly cited for the failure of the SFC was the collapse of crude oil prices during the 1980s, although other factors contributed.[91]Without a stable and predictable price for the commodity that the SFC was attempting to produce in specific, mandated quantities, the structure of the SFC was unable to cope with market changes:

> The failure of the federal government's effort to create a synthetic fuels industry yields valuable lessons about the role of government in technology innovation. The synthetic fuels program was established without sufficient flexibility to meet changes in market conditions, such as the price of fuel. Public unwillingness to endure the environmental costs of some of the large-scale projects was an added complication. An emphasis on production targets was an added complication. An emphasis on production targets reduced research and program flexibility. Rapid turnover among SFC's high-level officials slowed administrative actions. The synthetic fuels program did demonstrate, however, that large-scale synthetic energy projects could be build and operated within specified technical parameters.[92] [citations from original text omitted]

It may be argued that DOE's initial decision to "restructure" the FutureGen program (as originally conceived, see box above) was partly attributable to the project's inflexibility in dealing with changing market conditions, in this case the rise in materials and construction costs and the doubling of FutureGen's original price estimate. However, the analogy between FutureGen and the SFC is limited. Although the SFC failed in part because of collapsing oil prices (the costs of the SFC program could be measured against the benefits of producing oil), for FutureGen the value of CO_2 *avoided* (i.e., the benefit provided by the technology) is not even calculable for comparison to the costs of building the plant, because there is no real global price for CO_2.

The market conditions that contributed to the downfall of the SFC, however, could be very different from the market conditions that would arise following the creation of a price for CO_2 emissions. The stability and predictability of the price signal would depend on the mechanism: carbon tax, allowance pricing, or auctions. A mechanism that allowed for a long-term price signal for carbon would likely benefit CO_2 capture technology R&D programs.

IMPLICATIONS FOR CLIMATE CHANGE LEGISLATION

Any comprehensive approach to reducing greenhouse gases substantially must address the world's dependency on coal for one-quarter of its energy demand, including almost half of its electricity demand. To maintain coal as a key component in the world's energy mix in a carbon- constrained future would require developing a technology to capture and store its CO_2 emissions. This situation suggests to some that any greenhouse gas reduction program be delayed until suchcarbon capture technology has been demonstrated. However, technological innovation and the demands of a carbon control regime are interlinked; therefore, a technology policy is no substitute for environmental policy and must be developed in concert with it.[93]

This linkage raises issues for legislators attempting to craft greenhouse gas reduction legislation. For the demand-pull side of the equation, the issue revolves around how to create the appropriate market for emerging carbon capture technologies. **Table 4** compares four different "price" signals across five different criteria that influence their effectiveness in promoting technology:

- **Magnitude:** What size of price signal or stringency of the regulation is imposed initially?
- **Direction:** What influences the direction (up or down) of the price signal or stringency of the regulation over time?
- **Timing:** How quickly is the price or regulation imposed and strengthened?
- **Stability:** How stable is the price or regulation over time?
- **Duration:** How long is the price or regulation imposed on affected companies?

In general, the criteria suggest that regulation is the surest method of forcing the development of technology—price is not necessarily a direct consideration in decision-making. However, regulation is also the most limiting; technologies more or less stringent than the standard would have a limited domestic market (although foreign opportunities may be available), and development could be frozen if the standards are not reviewed and strengthened periodically.

Table 4. Comparison of Various Demand-Pull Mechanisms

Mechanism	Magnitude	Direction	Timing	Stability	Duration
Regulation	Depends on available technology or performance standard	Subject to periodic review by regulatory authorities based on technological progress	Depends on frequency of regulatory review and pace of technological progress	Very stable can become stagnant if discourages further inno-vation or regulators rarely review standard	Depends on the regulatory procedures for reassessment
Allowance Prices	Depends on stringency of emissions cap and other pro-visions of the cap-and-trade program	Market-driven based on the supply and demand for allowances	Depends on environmental goal and specified schedule of emission reductions	Can be quite volatile	Depends on environmental goal and speci-fied schedule of emission reductions
Carbon Tax	Depends on level of tax	Generally specified by legislation	Depends on escalator provisions in legislation	Stable	Depends on the specified schedule of the carbon tax
Allowance Auctions	Same dynamics as allowance prices; can be strengthened by 100% auctioning of allowances and specifying a reserve price	Same dynamics as allowance prices unless legislation specifies a reserve price	Same dynamics as allowance prices unless legislation includes a reserve price—then it depends on any escalator clause	Allowance price volat-ility can be tempered by a reserve price and the specifics of the auctio-ning process	Same as for allowance prices, but includes the details of the auctioning procedures

Source: Congressional Research Service.

In contrast, allowance prices would provide the most equivocal signal, particularly if they are allocated free to participants. Experience has shown allowance prices to be subject to volatility with swings both up and down. The experience with the SO_2 cap-and-trade program suggests the incentive can be improved with "bonus" allowances; however, the eligibility criteria used could be perceived as the government attempting to pick a winner.

In contrast, carbon taxes and allowance auctions (particularly 100% auctions with a reserve price) provide strong market-based price signals. A carbon tax is the most stable price signal, providing a clear and transparent signal of the value of any method of greenhouse gas reductions. Substantial auctioning of allowances also places a price on carbon emissions, a price that can be strengthened by incorporating a reserve price into the structure of the auction.

However, each of these signals ultimately depends on the environmental goal envisioned and the specifics of the control program: (1) the stringency of the reduction requirement; (2) the timing of desired reductions; (3) the techniques allowed to achieve compliance. The interplay of these factors informs the technology community about the urgency of the need for carbon capture technology; the price signal informs the community what cost-performance parameters are appropriate for the emerging carbon market. The nature of that price signal (regulatory, market, stability) informs the community of the confidence it can have that it is not wasting capital on a "white elephant" or on a project that the market does not want or need.

The issues for technology-push mechanisms are broader, and include not only the specifics of any reduction program and resulting price signal, but also international considerations and the interplay between carbon capture technology, storage, and the potential need for CO_2 transport. Groups as diverse as The Pew Center, the Electric Power Research Institute, DOE, and MIT have suggested "roadmaps" and other schemes for preparing carbon capture technology for a pending greenhouse gas reduction program.[94] Generally, all of these approaches agree on the need for demonstration-size (200-300 MW) projects to sort out technical performance and cost effectiveness, and identify potential environmental and safety concerns. The Energy Independence and Security Act of 2007 (P.L. 110-140) reflected Congress's desire for more integrated demonstration projects, and ARRA provides a dramatic step increase in funding for CCS technology over a relatively short period.

Finally, it should be noted that the status quo for coal with respect to climate change legislation isn't necessarily the same as "business as usual."

The financial markets and regulatory authorities appear to be hedging their bets on the outcomes of any federal legislation with respect to greenhouse gas reductions, and are becoming increasingly unwilling to accept the risk of a coal- fired power plant with or without carbon capture capacity. This sort of limbo for coal-fired power plants is reinforced by the MIT study, which makes a strong case against subsidizing new construction (allowed for IGCC under the EPAct2005) without carbon capture because of the unattractive costs of retrofits:

> *Coal plants will not be cheap to retrofit for CO_2 capture.* Our analysis confirms that the costs to retrofit an air-driven SCPC [supercritical pulverized coal] plant for significant CO_2 capture, say 90%, will be greater than the costs to retrofit an IGCC plant. However, ... the modifications needed to retrofit an IGCC plant for appreciable CCS are extensive and not a matter of simply adding a single simple and inexpensive process step to an existing IGCC plant.... Consequently, IGCC plants without CCS that receive assistance under the 2005 Energy Act will be more costly to retrofit and less likely to do so.
>
> *The concept of a "capture ready" IGCC or pulverized coal plant is as yet unproven and unlikely to be fruitful.* The Energy Act envisions "capture ready" to apply to gasification technology. [citation omitted] Retrofitting IGCC plants, or for that matter pulverized coal plants, to incorporate CCS technology involves substantial additional investments and a significant penalty to the efficiency and net electricity output of the plant. As a result, we are unconvinced that such financial assistance to conventional IGCC plants without CCS is wise.[95] [*emphasis in original*]

As noted earlier, lack of a regulatory scheme (or carbon price) presents numerous risks to any research and development effort designed to develop carbon capture technology. Ultimately, it also presents a risk to the future of coal.

End Notes

[1] Pew Center on Global Climate Change, *Coal and Climate Change Facts*, (2008), available at http://www.pewclimate.org/global-warming-basics/coalfacts.cfm.

[2] International Energy Agency, *World Energy Outlook 2007: China and India Insights* (2007), pp. 593.

[3] World, China and India statistics from International Energy Agency, *World Energy Outlook 2007: China and India Insights*, (2007), pp. 592, 596, and 600; United States statistics from Energy Information Administration, *Annual Energy Review: 2005* (July 2006), p. 228.

[4] Pew Center on Global Climate Change, *Coal and Climate Change Facts* (2008), available at http://www.pewclimate.org/global-warming-basics/coalfacts.cfm. Capacity factor derived by CRS from data presented, assuming plants would operate in baseload mode with 70% capacity factors.

[5] For a general discussion of carbon capture and sequestration, see CRS Report RL33801, *Carbon Capture and Sequestration (CCS)*, by Peter Folger.

[6] For more information on policy issues related to the transportation of CO_2, see CRS Report RL33971, *Carbon Dioxide (CO_2) Pipelines for Carbon Sequestration: Emerging Policy Issues*, and CRS Report RL343 16, *Pipelines for Carbon Dioxide (CO_2) Control: Network Needs and Cost Uncertainties*, both by Paul W. Parfomak and Peter Folger.

[7] For a useful summary of carbon capture technology, see Steve Blankinship, "The Evolution of Carbon Capture Technology Part 1," *Power Engineering* (March 2008).

[8] Ryan M. Dailey and Donald S. Shattuck, "An Introduction to CO_2 Capture and Sequestration Technology, *Utility Engineering*" (May 2008), p. 3.

[9] Massachusetts Institute of Technology, *The Future of Coal: An Interdisciplinary MIT Study* (2007), p. 147. Hereafter referred to as MIT, *The Future of Coal.*

[10] Levelized cost is the present value of the total cost of building and operating a generating plant over its economic life, converted to equal annual payments. Costs are levelized in real dollars (i.e., adjusted to remove the impact of inflation).

[11] MIT, *The Future of Coal*, pp. 27, 149.

[12] Ryan M. Dailey and Donald S. Shattuck, "An Introduction to CO_2 Capture and Sequestration Technology, *Utility Engineering*" (May 2008), p. 4.

[13] Satish Reddy, Dennis Johnson,, and John Gilmartin, *Fluror's Economine FG Plus*^{SM} *Technology for CO_2 Capture at Coal-fired Power Plants*, presented at Power Plant Air Pollutant Control "Mega" Symposium, Baltimore (August 25- 28, 2008).

[14] Heat-stable salts are acid anions with a stronger acid strength than the acid gases that are removed from the process gas. These anions may bind to the usable amine and then therefore make it unavailable for acid gas absorption; hence the need to inhibit their formulation or to purge them from the system.

[15] Satish Reddy, Dennis Johnson,, and John Gilmartin, *Fluror's Economine FG Plus*^{SM} *Technology for CO_2 Capture at Coal-fired Power Plants*, presented at Power Plant Air Pollutant Control "Mega" Symposium, Baltimore (August 25- 28, 2008), pp. 3-5.

[16] Slipstream refers to pilot testing at an operating power plant using a portion of the flue gas stream.

[17] HTC Pureenergy, *CO_2—Source to Sink*, presentation (April 2008). Also see webite at http://www.htcenergy.com.

[18] Ryan M. Dailey and Donald S. Shattuck, "An Introduction to CO_2 Capture and Sequestration Technology, *Utility Engineering*" (May 2008), p. 5.

[19] Steve Blankinship, "The Evolution of Carbon Capture Technology, Part 1," *Power Engineering* (March 2008), p. 30.

[20] AEP News Release, *RWE to Join AEP in Validation of Carbon Capture Technology*, (November 8, 2007).

[21] Ryan M Dailey and Donald S. Shattuck, "An Introduction to CO_2 Capture and Sequestration Technology, *Utility Engineering*" (May 2008), p. 7.

[22] Powerspan Corp., *Carbon Capture Technology for Existing and New Coal-Fired Power Plants* (April 15, 2008).

[23] Powerspan Corp., *Powerspan Announces CO_2 Capture Technology Pilot Test Results* (December 22, 2009).

[24] IGCC is an electric generating technology in which pulverized coal is not burned directly but mixed with oxygen and water in a high-pressure gasifier to make "syngas," a combustible fluid that is then burned in a conventional combined- cycle arrangement to generate power.

[25] MIT, *The Future of Coal*, p. 35.

[26] MIT, *The Future of Coal*, p. 36.

[27] Brad Kitchens and Greg Litra, "Restructuring FutureGen," *Electric Light & Power* (May/June 2008), pp. 46-47, 58.

[28] MIT, *The Future of Coal*, pp. 30-31.

[29] For more information, see Vattenfall's website at http://www.vattenfall.com/www/co$_2$_en/co$_2$_en/879177tbd/ 8792 1 1pilot/index.jsp

[30]For more information, see Callide Project website at http://www.callideoxyfuel.com/What/CallideOxyfuelProject.aspx.

[31] MIT, *The Future of Coal*, p. 147.

[32] MIT, *The Future of Coal*, pp. 30, 149.

[33] Steve Blankinship, "The Evolution of Carbon Capture Technology, Part 2," *Power Engineering* (May 2008), pp. 62-63.

[34] DOE National Energy Technology Laboratory, Carbon Sequestration FAQ Information Portal, at http://www.netl.doe.gov/technologies/carbon

[35] For example, see Interlaboratory Working Group, *Scenarios for a Clean Energy Future*, ORNL/CON-476 (November 2000).

[36] For example, see *CERA Advisory Service, Design Issues for Market-based Greenhouse Gas Reduction Strategies; Special Report* (February 2006), p. 59; Congressional Budget Office, *Evaluating the Role of Prices and R&D in Reducing Carbon Dioxide Emissions* (September 2006).

[37] Richard D. Morgenstern, *Climate Policy Instruments: The Case for the Safety Valve* (Council on Foreign Relations, September 20-21, 2004), p. 9.

[38] Carolyn Fischer, *Climate change Policy Choices and Technical Innovation*, Resources for the Future Issue Brief #20 (June 2000), p. 2.

[39] State Corporation Commission, *Application of Appalachian Power Company*, Case No. PUE-2007-00068 (Richmond, April 14, 2008), p. 16.

[40] As stated by DOE: "Regulatory uncertainty for GHG legislation is a key issue impacting technology selection and reliability of economic forecasts. Returns on investment for conventional plants, including supercritical, can be severely compromised by the need to subsequently address CO_2 mitigation. Higher capital costs incurred for IGCC may make such new plants less competitive unless their advantage in CO_2 mitigation is assured." DOE National Energy Technology Laboratory, *Tracking New Coal-fired Power Plants* (June 30, 2008), p. 14.

[41] Citi, Morgan Chase, and Morgan Stanley, *The Carbon Principles: Fossil Fuel Generation Financing Enhanced Environmental Diligence Process* (February 2008), p. 1.

[42] Exceptions to this would include areas where the carbon dioxide could be used for EOR, or where a state or region has enacted greenhouse gas controls, such as California and several northeastern states.

[43] For a fuller discussion of the uncertainties involved in estimating the cost of cap-and-trade programs, see CRS Report RL34489, *Climate Change: Costs and Benefits of S. 2191/S. 3036*, by Larry Parker and Brent D. Yacobucci.

[44] Segey Paltsev, et al., *Assessment of U.S. Cap-and-Trade Proposals*, MIT Joint Program on the Science and Policy of Global Change, Report 146 (April 2007), p. 16.

[45] The Clean Air Act, Section 111 (42 U.S.C. 7411).

[46] Donald Shattuck, et al., *A History of Flue Gas Desulfurization (FGD)—The Early Years*, UE Technical Paper (June 2007), p. 3.

[47] For a fuller discussion of EPA authority to regulate greenhouse gases under the Clean Air Act, see Robert J. Meyer, Principal Deputy Assistant Administrator, Office of Air and Radiation,

EPA Testimony before the Subcommittee on Energy and Air Quality, Committee on Energy and Commerce, U.S. House of Representation (April 10, 2008).

[48] For a discussion of the structure of the Clean Air Act, see CRS Report RL3 0853, *Clean Air Act: A Summary of the Act and Its Major Requirements*, by James E. McCarthy et al.

[49] 42 U.S.C. 7411, Clean Air Act, Sec. 11 1(a)(1)

[50] 40 CFR 60.40-46, Subpart D—Standards of Performance for Fossil-Fuel-Fired Steam Generator for Which Construction is Commenced After August 17, 1971.

[51] Margaret R. Taylor, *The Influence of Government Actions on Innovative Activities in the Development of Environmental Technologies to Control Sulfur Dioxide Emissions from Stationary Sources,* Thesis, Carnegie Institute of Technology (January 2001), p. 37, 40.

[52] Fri v. Sierra Club, 412 US 541 (1973). This decision resulted in EPA issuing "prevention of significant deterioration" regulations in 1974; regulations what were mostly codified in the 1977 Clean Air Amendment (Part C).

[53] Taylor, ibid., p. 37.

[54] Taylor, ibid., p. 39.

[55] For a discussion of challenges arising from the early development of FGD, see Donald Shattuck, et al., *A History of Flue Gas Desulfurization (FGD) —The Early Years*, UE Technical Paper (June 2007).

[56] Examples include full-page ads in the Washington Post entitled "Requiem for Scrubbers," "Scrubbers, Described, Examined and Rejected," and "Amen." For an example, see Washington Post, p. A32 (October 25, 1974).

[57] 40 CFR 60.40Da-52Da, Subpart Da—Standards of Performance for Electric Utility Stream Generating Units for Which Construction is Commenced After September 18, 1978.

[58] Margaret R. Taylor, Edward S. Rubin, and David A. Hounshell, "Control of SO2 Emissions from Power Plants: A Case of Induced Technological Innovation in the U.S.," *Technological Forecasting & Social Change* (July 2005), p. 697.

[59] Shattuck, et. al., p. 15.

[60] Congressional Budget Office, *Evaluating the Roles of Prices and R&D in Reducing Carbon Dioxide Emissions* (September 2006), p. 17.

[61] In addition, some of the revenue generated by the tax could be used to fund research, development, demonstration, and deployment of new technology to encourage the long-term transition to a less-carbon-intensive economy.

[62] For a fuller discussion, see CRS Report RL30853, *Clean Air Act: A Summary of the Act and Its Major Requirements*, by James E. McCarthy et al.

[63] Margaret R. Taylor, Edward S. Rubin, and David A. Hounshell, "Effect of Government Actions on Technological Innovation for SO2 Control," *Environmental Science & Technology* (October 15, 2003), p. 4531. In a more recent article, the authors state: "Finally, the case provides little evidence for the claim that cap-and-trade instruments induce innovation more effectively than other instruments." Margaret R. Taylor, Edward S. Rubin, and David A. Hounshell, "Control of SO2 Emissions from Power Plants: A Case of Induced Technological Innovation in the U.S.," *Technological Forecasting & Social Change* (July 2005), p. 697-8.

[64] Like a carbon tax, the revenues received could be at least partly directed toward research, development, and demonstration programs.

[65] Karsten Neuhoff, *Auctions for CO_2 Allowances—A Straw Man Proposal*, University of Cambridge Electricity Policy Research Group (May 2007), pp. 3-6. A short-squeeze is a situation where the price of a stock or commodity rises and investors who sold short (believing the price was going to fall) rush to buy it to cover their short position and cut their losses.

[66] For more information on the emission allowance schemes for CCS in legislation, see CRS Report R40867, *Carbon Capture and Sequestration in H.R. 2454 and S. 1733*, by Peter Folger, Mary Tiemann, and Stan Mark Kaplan.

[67] Personal communication, Timothy E. Fout, General Engineer, DOE National Energy Technology Laboratory, Morgantown, WV (July 16, 2008).

[68] Funding for FY2009 is according to U.S. Department of Energy, *FY2011 Congressional Budget Request*, volume 3, Fossil Energy Research and Development, at http://www.cfo.doe.gov/budget/11budget/Content/Volume%203.pdf.

[69] DOE FY2011 Congressional Budget Request, p. 649, at http://www.cfo.doe.gov/budget/ 11budget/Content/ Volume%203.pdf.

[70] Ibid.

[71] Approximately $1.4 billion through FY2009 (CRS estimate).

[72] U.S. Congress, House Committee on Appropriations, *Omnibus Appropriations Act, 2009, Division C—Energy and Water Development and Related Agencies Appropriations Act, 2009*, committee print, 111th Cong., 1st sess., March 11, 2009, p. 672.

[73] See quote by Morgenstern above. In that analysis, government-supported research is needed to compensate for market imperfections. In the current situation, there is no market, and thus its imperfections are moot.

[74] MIT, *The Future of Coal*, p. xi.

[75] Carolyn Fischer, *Climate Change Policy Choices and Technical Innovation*, Resources for the Future Climate Issue Brief #20 (June 2000), p. 9

[76] A distribution utility is defined in the legislation as an electric utility that has a legal, regulatory, or contractual obligation to deliver electricity directly to retail customers.

[77] MIT, *The Future of Coal*, p. 102.

[78] S. 3132 would also provide a $10 per metric ton credit for CO_2 captured and used as a tertiary injectant in an enhanced oil and natural gas recovery project.

[79] For a comparison between both bills, see CRS Report R40867, *Carbon Capture and Sequestration in H.R. 2454 and S. 1733*, by Peter Folger, Mary Tiemann, and Stan Mark Kaplan.

[80] For example, H.R. 2809, the New Apollo Energy Act of 2007; and H.R. 6385, the Apollo Energy Independence Act of 2008.

[81] For more information on this topic, see CRS Report RL34645, *The Manhattan Project, the Apollo Program, and Federal Energy Technology R&D Programs: A Comparative Analysis*, by Deborah D. Stine.

[82] U.S. Department of Energy, Office of History and Heritage Resources, "The Manhattan Project: An Interactive History," webpage at http://www.cfo.doe.gov/me70/manhattan/ 1939-1942.htm. F.G. Gosling, *The Manhattan Project: Making the Atomic Bomb*, January 1999 edition (Oak Ridge, TN: Department of Energy).

[83] Richard G. Hewlett and Oscar E. Anderson, Jr., A *History of the United States Atomic Energy Commission: The New World, 1939/1946 ,Volume I,* (University Park, PA: The Pennsylvania State University Press, 1962). Appendix 2 provides the annual Manhattan project expenditures. These costs were adjusted to 2007 dollars using the price index for gross domestic product (GDP), available from the Bureau of Economic Affairs, National Income and Product Accounts Table webpage, Table 1.1.4., at http://www. bea.gov/bea/dn/nipaweb/.

[84] There is some difference of opinion regarding what activities comprise the Apollo program, and thus when it begins and ends. Some include the first studies for Apollo, Skylab, and the use of Apollo spacecraft in the Apollo-Soyuz Test Project. This analysis is based on that provided by the National Aeronautics and Space Administration (NASA), which includes the first studies of Apollo, but not Skylab or Soyuz activities, in a 2004 web update by Richard Orloff of its publication entitled *Apollo By The Numbers: A Statistical Reference*, NASA SP-2000-4029, at http://history.nasa.gov/ SP-4029/Apollo_00_Welcome.htm.

[85] The funding data is available at http://history.nasa.gov/SP-4214/app2.html#1965. It is based on information in NASA, *The Apollo Spacecraft - A Chronology*, NASA Special Publication-4009, at http://www.hq.nasa.gov/office/pao/ History/SP-4009/contents.htm. This data is

from Volume 4, Appendix 7 at http://www.hq.nasa.gov/office/pao/History/ SP-4009/v4app7.htm.

[86] Richard Orloff, *Apollo By The Numbers: A Statistical Reference*, NASA SP-2000-4029, 2004 web update, at http://history.nasa.gov/SP-4029/Apollo_00_Welcome.htm. The funding data is available at http://history.nasa.gov/SP4029/Apollo_1 8-16_Apollo_Program_Budget_ Appropriations.htm. It is based on information in NASA, *The Apollo Spacecraft - A Chronology*, NASA Special Publication-4009, at http://www.hq.nasa.gov/office/ pao/History/SP-4009/ contents.htm.

[87] Department of Energy, "Origins & Evolution of the Department of Energy," webpage at http://www.doe.gov/about/ origins.htm.

[88] The Department of Energy Organization Act of 1977 (P.L. 95-91).

[89] CRS estimates that federal funding to address global climate change was $6.37 billion in FY2008. See CRS Report RL33817, *Climate Change: Federal Program Funding and Tax Incentives*, by Jane A. Leggett.

[90] The National Academy of Sciences, "The Government Role in Civilian Technology: Building a New Alliance" (National Academy Press, Washington, DC, 1992), pp. 58-59.

[91] For a variety of reasons, Canada's experience with producing synthetic fuels, specifically oil sands development, has differed from the U.S. experience. For more information, see CRS Report RL34258, *North American Oil Sands: History of Development, Prospects for the Future*, by Marc Humphries.

[92] The National Academy of Sciences, "The Government Role in Civilian Technology: Building a New Alliance," p. 59.

[93] Carolyn Fischer, *Climate Change Policy Choices and Technical Innovation*, Resources for the Future Climate Issue Brief #20 (June 2000), p. 9.

[94] For example, see Pew Center on Global Climate Change, *Coal and Climate Change Facts*, (2008), available at http://www.pewclimate.org/global-warming-basics/coalfacts.cfm; Coal Utilization Research Council and Electric Power Research Institute technology roadmap at http://www.coal DOE Energy, National Energy Technology Laboratory, *Carbon Sequestration Technology Roadmap and Program Plan 2007* available at http://www.netl.doe.gov/technologies/carbon and, MIT, *The Future of Coal*, pp. xi-xv.

[95] MIT, *The Future of Coal*, pp. 98-99.

In: Coal-Fired Power Plants and Carbon... ISBN: 978-1-61122-054-4
Editor: Matthew W. Shuester

Chapter 2

Coal Power Plants: Opportunities Exist for DOE to Provide Better Information on the Maturity of Key Technologies to Reduce Carbon Dioxide Emissions

United States Government Accountability Office

Why GAO Did This Study

Coal power plants generate about half of the United States' electricity and are expected to remain a key energy source. Coal power plants also account for about one-third of the nation's emissions of carbon dioxide (CO_2), the primary greenhouse gas that experts believe contributes to climate change. Current regulatory efforts and proposed legislation that seek to reduce CO_2 emissions could affect coal power plants. Two key technologies show potential for reducing CO_2 emissions: (1) carbon capture and storage (CCS), which involves capturing and storing CO_2 in geologic formations, and (2) plant efficiency improvements that allow plants to use less coal.

The Department of Energy (DOE) plays a key role in accelerating the commercial availability of these technologies and devoted more than $600 million to them in fiscal year 2009. Congress asked GAO to examine (1) the maturity of these technologies; (2) their potential for commercial use, and any challenges to their use; and (3) possible implications of deploying these technologies. To conduct this work, GAO reviewed reports and interviewed stakeholders with expertise in coal technologies.

What GAO Recommends

GAO recommends that DOE develop a standard set of benchmarks to gauge and report to Congress on the maturity of key technologies. In commenting on a draft of this report, DOE concurred with our recommendation.

What GAO Found

DOE does not systematically assess the maturity of key coal technologies, but GAO found consensus among stakeholders that CCS is less mature than efficiency technologies. Specifically, DOE does not use a standard set of benchmarks or terms to describe the maturity of technologies, limiting its ability to provide key information to Congress, utilities, and other stakeholders. This lack of information limits congressional oversight of DOE's expenditures on these efforts, and it hampers policymakers' efforts to gauge the maturity of these technologies as they consider climate change policies. In the absence of this information from DOE, GAO interviewed stakeholders with expertise in CCS or efficiency technologies to identify their views on the maturity of these technologies. Stakeholders told GAO that while components of CCS have been used commercially in other industries, their application remains at a small scale in coal power plants, with only one fully integrated CCS project operating at a coal plant. Efficiency technologies, on the other hand, are in wider commercial use.

Commercial deployment of CCS is possible within 10 to 15 years while many efficiency technologies have been used and are available for use now. Use of both technologies is, however, contingent on overcoming a variety of economic, technical, and legal challenges. In particular, with respect to CCS, stakeholders highlighted the large costs to install and operate current CCS technologies, the fact that large scale demonstration of CCS is needed in coal plants, and the lack of a national carbon policy to reduce CO_2 emissions or a legal framework to govern liability for the permanent storage of large amounts of CO_2. With respect to efficiency improvements, stakeholders highlighted the high cost to build or upgrade such coal plants, the fact that some upgrades require highly technical materials, and plant operators' concerns that changes to the existing fleet of coal power plants could trigger additional regulatory requirements.

CCS technologies offer more potential to reduce CO_2 emissions than efficiency improvements alone, and both could raise electricity costs and have other effects. According to reports and stakeholders, the successful deployment of CCS technologies is critical to meeting the ambitious emissions reductions that are currently being considered in the United States while retaining coal as a fuel source. Most stakeholders told GAO that CCS would increase electricity costs, and some reports estimate that current CCS technologies would increase electricity costs by about 30 to 80 percent at plants using these technologies. DOE has also reported that CCS could increase water consumption at power plants. Efficiency improvements offer more potential for near term reductions in CO_2 emissions, but they cannot reduce CO_2 emissions from a coal plant to the same extent as CCS.

ABBREVIATIONS

ARRA	American Recovery and Reinvestment Act
CCPI	Clean Coal Power Initiative
CCS	carbon capture and storage
CO	carbon monoxide
CO_2	carbon dioxide
DOD	Department of Defense
DOE	Department of Energy
EIA	Energy Information Administration
EOR	enhanced oil recovery
EPA	Environmental Protection Agency
EPRI	Electric Power Research Institute
GHG	greenhouse gas
IEA	International Energy Agency
IGCC	Integrated Gasification Combined Cycle
IPCC	Intergovernmental Panel on Climate Change
MIT	Massachusetts Institute of Technology
MW	megawatt

NAS	National Academy of Sciences
NASA	National Aeronautics and Space Administration
NERC	North American Electric Reliability Corporation
NSR	New Source Review
RD&D	research, development, and demonstration
SDWA	Safe Drinking Water Act
TRL	Technology Readiness Level

June 16, 2010

The Honorable James M. Inhofe
Ranking Member
Committee on Environment and Public Works
United States Senate

The Honorable George V. Voinovich
United States Senate

Coal power plants generate about half of the United States' electricity and are expected to continue supplying a large portion of the nation's electricity in the future. According to the Department of Energy's (DOE) Energy Information Administration (EIA),[1] coal will provide 44 percent of the electricity in 2035 in the United States. The critical role that coal plays in supplying electricity is due in part to the large coal reserves in the United States, which some estimate will last about 240 years at current consumption levels, and the relatively low cost of this energy supply. However, coal power plants also currently account for about one-third of the nation's emissions of carbon dioxide (CO_2), the most prevalent greenhouse gas. Concerns over rising greenhouse gas emissions and their potential effects on the climate have led some countries to adopt or consider adopting policies to reduce these emissions. In the United States and elsewhere, these concerns have also increased focus on developing and using technologies to limit CO_2 emissions from coal power plants while allowing coal to remain a viable source of energy.

Two key technologies show potential for reducing CO_2 emissions from coal plants: carbon capture and storage (CCS) and efficiency technologies. CCS technologies separate and capture CO_2 from other gases produced when combusting or gasifying coal, compress it, then transport it to underground geologic formations such as saline aquifers—porous rock filled with brine—

where it is injected for long-term storage. There are three approaches to capturing CO_2—post-combustion, pre-combustion, and oxy-combustion. Post-combustion capture involves capturing CO_2 from the exhaust stream created when coal is burned at pulverized coal plants, which make up nearly all coal plants operating in the United States.

Pre-combustion capture involves capturing CO_2 produced when gasifying coal at Integrated Gasification Combined Cycle (IGCC) plants, which are in limited use in the electricity industry. Oxy-combustion capture involves capturing CO_2 from the exhaust stream created when coal is burned in an oxygen-enriched environment at pulverized coal plants.

Efficiency technologies include more efficient designs for new coal power plants—such as IGCC plants, as well as ultrasupercritical plants—that operate at higher steam temperatures and pressures than conventional plants.[2] Efficiency upgrades can also be made in existing coal plants, such as overhauling or replacing turbine fan blades. Improving the efficiency of coal plants allows them to use less coal per unit of electricity produced and achieve a corresponding reduction in CO_2 emissions. CCS technologies and efficiency technologies can be used independently or in conjunction with one another.

In the United States, regulatory efforts and proposed legislation that seek to reduce CO_2 emissions could affect coal power plants. The Environmental Protection Agency (EPA) has taken steps to regulate greenhouse gas emissions under the Clean Air Act and plans to begin regulating emissions from certain stationary sources, including coal power plants, beginning in 2011. As part of this effort, EPA is compiling technical and background information on potential control technologies and measures, such as CCS, and developing policy guidance to assist permitting agencies in determining the best available control technology for greenhouse gas emissions. In addition, the American Clean Energy and Security Act passed the House of Representatives on June 26, 2009, and would require an 83 percent reduction in greenhouse gas emissions from 2005 levels by 2050.[3] Among other things, this proposed legislation would create a cap and trade program, a market-based mechanism to establish a price for emissions of greenhouse gases, and require additional specific actions to reduce these emissions. For example, section 116 of the bill would require new coal power plants permitted before 2020 to reduce their CO_2 emissions by half, 4 years after certain CCS deployment criteria are met or by 2025, whichever comes first.[4]

DOE plays a key role in accelerating the commercial availability of technologies to reduce CO_2 emissions from coal power plants. Specifically, DOE's Office of Fossil Energy oversees research on these technologies through its coal research, development, and demonstration (RD&D) program. This program carries out three primary activities: (1) managing and performing energy-related research that reduce barriers to the environmentally sound use of fossil fuels, (2) partnering with industry to advance technologies toward commercialization, and (3) supporting the development of information and policy options that benefit the public. Such information could help EPA in its review of available technologies to reduce CO_2 from coal plants along with other policymakers that are considering climate change policies. In the near term, according to DOE's fiscal year 2011 budget submission, DOE hopes to facilitate the development of CCS and efficiency technologies, with longer term goals of improving these technologies so that coal can remain part of the nation's fuel mix in generating electricity. In fiscal year 2009, DOE's coal RD&D funding was at least $681 million, and $3.4 billion was appropriated in the American Recovery and Reinvestment Act of 2009 (ARRA) for fossil energy RD&D.[5]

In this context, you asked us to review key technologies to reduce CO_2 emissions from coal power plants. Specifically, we examined (1) the maturity of technologies to reduce CO_2 emissions from coal power plants; (2) the potential for these technologies to be used commercially in the future, and challenges, if any, to their use; and (3) the possible implications of deploying these technologies. We briefed your staffs on the results of our work on June 1, 2010 (see app. I). This report summarizes and formally transmits the information provided during that briefing. It incorporates technical and other comments provided by agencies since the briefing.

To conduct this work, we reviewed key reports including those from DOE's national laboratories, the National Academy of Sciences (NAS), International Energy Agency (IEA), Intergovernmental Panel on Climate Change (IPCC), Global CCS Institute, the National Coal Council, and academic reports. We conducted interviews with stakeholders such as power plant operators, technology vendors, and federal officials from EPA and DOE along with officials from the North American Electric Reliability Corporation (NERC). We then selected a group of 19 stakeholders with expertise in CCS or efficiency technologies to answer a standard set of questions. This group included those from major utilities that are planning or implementing projects using key coal technologies, technology vendors that are developing these technologies, federal officials providing RD&D funding for these technologies, and researchers from academia and industry that are researching these technologies. We asked these stakeholders to describe the maturity of these technologies using a nine point scale we developed in conjunction with the Electric Power Research Institute (EPRI) based on the National Aeronautics and Space Administration's (NASA) Technology Readiness Levels (TRL).[6] TRLs are a tool that is used by NASA and other agencies to rate the extent to which technologies have been demonstrated to work as intended. We also reviewed available data on the use of key coal technologies compiled by IEA and the Global CCS Institute.

To identify the potential for these technologies to be used commercially in the future along with any associated challenges or implications, we reviewed key reports on CCS and efficiency technologies and examined goals set out by DOE, IEA, and electricity industry groups for deploying these technologies. We also asked our 19 stakeholders with expertise in CCS or efficiency technologies for their views on the potential challenges and implications of using these technologies. Finally, we visited coal power plants and research facilities in three states—Alabama, Maryland, and West Virginia—that we selected because they contained projects involving advanced coal technologies. Importantly, our discussion focuses on the technological maturity of these technologies. TRLs describe the level of demonstration achieved for particular technologies, but they do not provide information on other factors that play a critical role in decisions to deploy them, such as their cost, availability of financing, and applicable regulations. Technological improvements could help these technologies overcome some challenges or potential negative implications. For example, novel approaches to CO_2 capture could help to lower the cost of using these technologies.

We conducted this performance audit from July 2009 through May 2010 in accordance with generally accepted government auditing standards. Those standards require that we plan and perform the audit to obtain sufficient, appropriate evidence to provide a reasonable basis for our findings and conclusions based on our audit objectives. We believe that the evidence obtained provides a reasonable basis for our findings and conclusions based on our audit objectives. A more detailed description of our scope and methodology is presented in appendix II.

ALTHOUGH DOE DOES NOT SYSTEMATICALLY ASSESS THE MATURITY OF KEY COAL TECHNOLOGIES, CONSENSUS AMONG STAKEHOLDERS IS THAT CCS IS LESS MATURE THAN EFFICIENCY TECHNOLOGIES

DOE's Office of Fossil Energy oversees research on key coal technologies, but DOE does not systematically assess the maturity of those technologies. Using TRLs we developed for these technologies, we found consensus among stakeholders that CCS is less mature than efficiency technologies.

DOE Does Not Systematically Assess the Maturity of Key Coal Technologies

Although federal standards for internal control require agency managers to compare actual program performance to planned or expected results and analyze significant differences,[7] we found that DOE's Office of Fossil Energy does not systematically assess the maturity of key coal technologies as they progress toward commercialization. While DOE officials reported that individual programs are aware of the maturity of technologies and DOE publishes reports that assess the technical and economic feasibility of advanced coal technologies, we found that the Office of Fossil Energy does not use a standard set of benchmarks or terms to describe or report on the maturity of technologies. In addition, DOE's goals for advancing these technologies sometimes use terms that are not well defined. The lack of such benchmarks or an assessment of the maturity of key coal technologies and whether they are achieving planned or desired results limits:

- DOE's ability to provide a clear picture of the maturity of these technologies to policymakers, utilities officials, and others;
- congressional and other oversight of the hundreds of millions of dollars DOE is spending on these technologies; and
- policymakers' ability to assess the maturity of CCS and the resources that might be needed to achieve commercial deployment.

Table 1. NASA's Technology Readiness Levels

TRL	Summary of TRL descriptions used by NASA
9	Actual system "flight proven" through successful mission operations under operational mission conditions
8	Actual system completed and "flight qualified" through test and demonstration. Examples include test and evaluation of the system in its intended weapons system to see if it meets design specifications
7	System prototype demonstration in realistic environment. Requires demonstration of actual system prototype in a realistic environment, such as an aircraft vehicle or space
6	System/subsystem model or prototype demonstration in a relevant environment
5	Component and/or breadboard validation in a relevant environment, which could be lab or simulated realistic environment
4	Component and/or breadboard validation in lab environment
3	Proof of concept test in lab environment
2	Technology concept and/or application formulated
1	Basic principles observed and reported

Source: GAO analysis of NASA data.

Other agencies similarly charged with developing technologies, such as NASA and the Department of Defense (DOD), use TRLs to characterize the maturity of technologies.[8] Table 1 shows a description of TRLs used by NASA.

DOE has acknowledged that TRLs can play a key role in assessing the maturity of technologies during the contracting process. The agency recently issued a *Technology Readiness Assessment Guide*, which lays out three key steps to conducting technology readiness assessments during the contracting process.[9]

- Identify critical technology elements that are essential to the successful operation of the facility.
- Assess maturity of these critical technologies using TRLs.
- Develop a technology maturity plan which identifies activities required to bring technology to desired TRL level.

Table 2. Scale Used to Gauge the Maturity of Coal Technologies

TRL	Description of TRLs we developed for coal technologies
9	Commercial operation in relevant environment (500 megawatt [MW] coal plant or greater, or about 3 million tons of CO_2 captured, transported, or stored annually)
8	Demonstration at more than 5 percent commercial scale (at least 125 MW coal captured, transported, or stored annually)
7	Pilot plant at more than about 5 percent commercial scale (at least 20 MW coal plant, or 100,000 tons of CO_2 captured, transported, or stored annually)
6	Process development unit at between about 0.1 percent to 5 percent of commercial scale (between 0.5 MW and 20 MW coal plant, or between about 3,000 and 100,000 tons of CO_2captured, transported, or stored annually)
5	Component validation in relevant environment (coal plant)
4	Component tests in lab
3	Proof of concept test
2	Application formulated (on paper)
1	Basic principles observed

Source: GAO framework analysis based on adaptation of TRLs to coal power plants and conversations with EPRI officials.

Note: We described commercial scale coal plant as 500 MW that emits 3 million tons of CO_2 annually. This is the size of a plant that has been used as a reference plant in engineering studies. Actual emissions from a coal plant can vary based on a variety of factors, including how often a plant operates.

Although use of the *Guide* is not mandatory, DOE's Office of Environmental Management uses the *Guide* as part of managing its procurement activities—a result of a GAO recommendation—and its Office of Nuclear Energy has begun using TRLs to measure and communicate risks associated with using critical technologies in a novel way.[10] Furthermore, the National Nuclear Security Administration has used TRLs recently as well.

Consensus among Key Stakeholders Is That CCS Is Less Mature than Efficiency Technologies

In the absence of an assessment from DOE, we asked stakeholders to gauge the maturity of coal technologies using a scale we developed based on TRLs. Table 2 shows the TRLs we developed for coal technologies by adapting the NASA TRLs.

Using the scale we developed for coal technologies, the consensus among key stakeholders we spoke with is that CCS is less mature than efficiency technologies. While all of the components of CCS—CO_2 capture, transportation, and storage—have been used commercially in other industries, such as natural gas processing and oil production, stakeholders generally reported that the application of these technologies remains at small scale in coal plants. Using TRLs, stakeholders generally reported that the largest demonstration of carbon capture in a coal plant was at a pilot scale (TRL 7) or less. Moreover, stakeholders identified only one integrated CCS system in a coal power plant—the Mountaineer Plant in West Virginia—which aims to capture and store more than 100,000 tons of CO_2.[11] This project captures CO_2 from a portion of the plant's exhaust—20 MW or about 4 percent the size of a typical 500 MW coal plant. DOE has announced funding for several integrated CCS projects in coal plants at larger scales—60 to 450 MW. In contrast to CCS, stakeholders generally told us that technologies that improve the efficiency of new or existing plants have already been demonstrated commercially. For example, a number of ultrasupercritical plants ranging from 600 to more than 1,000 MW have been built or are under construction in Europe and Asia, and there are five IGCC plants in operation around the world, including two in the United States.[12]

Commercial Deployment of Key Coal Technologies Is Possible, but Contingent on Overcoming Economic, Technical, and Legal Challenges

Commercial deployment of CCS within 10 to 15 years is possible according to DOE and other stakeholders, but is contingent on overcoming a variety of economic, technical, and legal challenges.[13] Many technologies to improve plant efficiency have been used and are available for commercial use now, but still face challenges.

Commercial Deployment of CCS Is Possible within 10 to 15 Years, but Faces Major Challenges According to Reports and Stakeholders

While DOE, electric industry groups, and other stakeholders have set goals to commercially deploy CCS in coal plants in the next 10 to 15 years, they acknowledge that these goals present significant challenges. In particular, they have highlighted the large costs to install and operate current CCS technologies. In 2007, DOE estimated the cost to install current CCS technologies was 85 percent higher for plants with post-combustion capture and was 36 percent higher for pre-combustion capture at IGCC plants, compared to comparable plants without CCS.[14] In addition, the large amount of energy that current CCS technologies require to operate— known as parasitic load—reduces the electricity plants can sell and raises operating costs. Parasitic load is estimated to be between 21 percent and 32 percent of plant output for post-combustion CO_2 capture and between 15 percent and 22 percent for pre-combustion CO_2 capture. To help reduce parasitic load of current technologies, DOE is supporting research on more advanced capture processes, including post-combustion work on membranes to capture CO_2 that may lower the cost of the current method of using chemical solvents.

In addition, key studies report that demonstration of large scale integrated CCS systems is a technical challenge and is needed to demonstrate the performance and potential costs of these systems. Some stakeholders also reported that additional demonstration was needed to lower perceived risk of technologies. For example, officials from one large utility told us that demonstration projects were needed to build experience with the technologies and to build vendor confidence so that they could provide technology performance guarantees. Similarly, officials from one state public utility commission reported that they considered CCS immature and were unlikely to approve cost recovery for such a project in the foreseeable future. Officials from two financial firms reported that they considered the application of CCS technologies at coal plants largely unproven and they would require additional demonstration projects or technology cost and performance guarantees from vendors or utilities to reduce the risk of financing these types of projects.

Moreover, without a national carbon policy to reduce CO_2 emissions nearly all stakeholders said CCS would not be widely deployed. Without a tax or a sufficiently restrictive limit on CO_2 emissions, plant operators lack an economic incentive to use CCS technologies. Reports by IPCC, NAS, and the Global CCS Institute have all highlighted the importance of a carbon policy to incentivize the use of CCS. In addition, nearly all stakeholders cited as challenges the lack of a regulatory framework to govern the permanent storage of large amounts of CO_2 in saline formations and legal uncertainty regarding long-term liability for the storage of CO_2. In 2008, EPA proposed a rule for injection of CO_2 for geologic sequestration under the Safe Drinking Water Act (SDWA).[15] EPA has stated that it lacks authority to release CO_2 injection well operators from liability for endangerment of underground sources of drinking water until the operator meets all the closure and post-closure requirements and EPA approves site closure of the well. According to EPA, once site closure is approved, well operators will only be liable under the SDWA if they violate or fail to comply with EPA orders in situations where an imminent and substantial endangerment to health is posed by a contaminant that is in or likely to enter an underground source of drinking water.[16] EPA plans to finalize the geologic sequestration rule in fall 2010. Neither the proposed rule nor the final rule will address liability for unintended releases of stored CO_2 that have other harmful effects. However, potential storage site operators are unlikely to assume these risks.

Many Efficiency Technologies Have Been Used and Are Ready for Commercial Use Now, but Also Face Challenges

Several stakeholders told us that building ultrasupercritical or IGCC plants may not be cost-effective for power plant owners in the United States because low coal prices limit the incentive to build highly efficient, but more costly, plants. Ultrasupercritical plants have higher capital costs because they use advanced materials, which may not justify expected fuel savings. To date, all of the more efficient ultrasupercritical plants have been built outside the United States, where coal prices are generally higher. Similarly, IGCC plants are more expensive than traditional pulverized coal units. According to some stakeholders, if low natural gas prices persist, utilities may choose to build natural gas power plants to reduce CO_2 emissions in lieu of more efficient coal plants.

In addition, some higher efficiency plant designs also face technical challenges in that they require more advanced materials than are currently available. For example, "advanced" ultrasupercritical plants require development of metal alloys to withstand steam temperatures that could be 300 to 500 degrees Fahrenheit higher than today's ultrasupercritical plants according to DOE.[17] From a legal perspective, most stakeholders reported that making efficiency upgrades to the existing fleet of coal power plants was limited by the prospect of triggering the Clean Air Act's New Source Review (NSR) requirements—additional requirements that may apply when a plant makes a major modification, a physical or operational change that would result in a significant net increase in emissions.

CCS Offers More Potential to Reduce CO_2 Emissions than Efficiency Improvements Alone; Both Could Have Cost and Other Effects

CCS technologies offer more potential to reduce CO_2 emissions than efficiency improvements alone but could raise electricity costs, increase demand for water, and could affect the ability of individual plants to operate reliably. Technologies to improve plant efficiency offer potential near-term reductions, but also raise some concerns.

CCS Could Help Meet Emissions Limits but Raises Key Concerns

According to key reports and stakeholders, the successful deployment of CCS technologies is critical to helping the United States meet potential limits in greenhouse gas emissions. In addition, CCS could allow coal to remain part of the nation's diverse fuel mix. IEA estimated that CCS technologies could meet 20 percent of reductions needed to reduce global CO_2 emissions by half by 2050.[18]This report also noted that the cost of meeting this goal would increase if CCS was not deployed. Massachusetts Institute of Technology (MIT) researchers called CCS the "critical enabling technology" to reduce CO_2 emissions while allowing continued use of coal in the future.[19] In 2009, NAS reported that if CCS technologies are not demonstrated commercially in the next decade, the electricity sector could move more towards using natural gas to meet emissions targets.[20] Our past work has also found that switching from coal to natural gas can lead to higher fuel costs and increased exposure to the greater price volatility of natural gas.[21]

On the other hand, most stakeholders told us that CCS would increase electricity prices, and key reports raise similar concerns. MIT estimated that plants with post-combustion capture have 61 percent higher cost of electricity, and IGCC plants with pre-combustion capture have a 27 percent higher cost compared to plants without these technologies.[22] Similarly, DOE estimated that plants with post-combustion capture have 83 percent higher cost of electricity, while IGCC plants with precombustion capture having a 36 percent higher cost.[23] DOE has also raised concerns about CCS and water consumption. Specifically, DOE estimated that post-combustion capture technology could almost double water consumption at a coal plant, while pre-combustion capture would increase water use by 73 percent.[24] Some utility officials also said CCS could lead to a decline in the ability of individual plants to operate reliably because a power plant might need to shut down if any of the three components (capture, transport, and storage) of CCS became unavailable. In addition, more electricity sources would need to make up for the higher parasitic load associated with CCS. The National Coal Council has also reported temporary declines in reliability during past deployments of new coal technologies.[25]

Plant Efficiency Improvements Offer More Potential for Near-Term Emissions Reductions but Also Raise Concerns

Because they have been used commercially already, technologies that improve plant efficiency offer the potential for near term reductions in CO_2 emissions. For example, DOE has estimated that efficiency improvements to the existing coal fleet could reduce CO_2 emissions by 100 million tons annually, or about a 5 to 10 percent reduction in overall emissions from these plants. According to the National Coal Council, increasing efficiency is the "only practical method for mitigating CO_2 emissions now" in coal plants.[26]

However, there are limits in the amount of CO_2 reductions that efficiency technologies can achieve. An ultrasupercritical plant emits about one-third less CO_2 than an average plant in the United States. By comparison, CCS offers the potential to capture 90 percent of a plant's CO_2 emissions. DOE officials and other stakeholders told us that plant efficiency improvements alone cannot reduce the CO_2 emissions from a coal plant to the same extent as CCS.

However, plant efficiency improvements can help to facilitate CCS because they reduce the amount of CO_2 that must be handled by the system. Finally, stakeholders' views were mixed on the potential effect of efficiency technologies on electricity costs, but they generally did not think efficiency technologies would increase water demand or compromise reliability.

CONCLUSIONS

Addressing climate change while retaining the use of coal to generate electricity will likely require the successful deployment of CCS and efficiency technologies in coal power plants. CCS, in particular, remains relatively immature compared to efficiency technologies, but offers the potential to reduce CO_2 emissions from power plants to a greater extent. The current regulatory and legislative efforts to reduce CO_2 emissions at coal power plants include consideration of the commercial availability of CCS. DOE plays a key role both in its efforts to advance CCS and efficiency technologies toward commercialization and in giving policymakers an accurate view of their maturity. However, because the agency does not systematically assess their development, DOE is unable to provide a clear picture of the maturity of these technologies or the necessary resources that might be required to move these technologies toward commercial demonstration. This lack of information limits congressional oversight of the hundreds of millions of dollars DOE is currently spending annually on efforts to advance coal technologies, and it hampers policymakers' efforts to gauge the maturity of these technologies as they consider climate change policies.

RECOMMENDATION FOR EXECUTIVE ACTION

To improve decision making and oversight for coal research efforts, including how technological maturity is measured and reported, we are making one recommendation to the Secretary of Energy. We recommend that the Secretary of Energy direct the Office of Fossil Energy to develop a standard set of benchmarks to gauge the maturity of key technologies and report to Congress on the maturity of these technologies. As part of this process, the Office of Fossil Energy should consider consulting DOE's

Technology Readiness Assessment Guide to develop benchmarks and reporting requirements.

APPENDIX I. BRIEFING SLIDES TO CONGRESSIONAL STAFF

Technologies to Reduce Carbon Dioxide Emissions from Coal Power Plants

Briefing to the Senate Committee on Environment and Public Works June 1, 2010

Introduction

Coal Plays Key Role in U.S. Electricity Sector but Emits Large Amount of Carbon Dioxide (CO_2)

Coal power plants

- provide about half of U.S. electricity (see figure 1)
- provide over 90% of electricity generated in some states
- account for about one-third of all U.S. emissions of CO_2

CO_2 is the most prevalent greenhouse gas (GHG)

- Concerns over rising GHG emissions and their potential effects on climate have led some countries to adopt or consider adopting policies to reduce these emissions

Two Key Technologies Show Potential for Reducing CO_2 Emissions from Coal Plants

Carbon capture and storage (CCS) is one of two key technologies for reducing CO_2 emissions from coal plants

- CO_2 is captured in one of three ways (see figures. 2, 3, and 4)
    - Post-combustion
    - Pre-combustion

- Oxy-combustion
- Captured CO_2 is compressed and transported via pipelines to underground geologic formations, where it is injected for long term storage, also known as sequestration

Integrated CCS projects involve all of these components: CO_2 capture, compression, transportation, and storage

Post-combustion captures CO_2 produced when burning coal in air

- Compatible with traditional pulverized coal plants, which make up nearly all coal plants currently operating worldwide

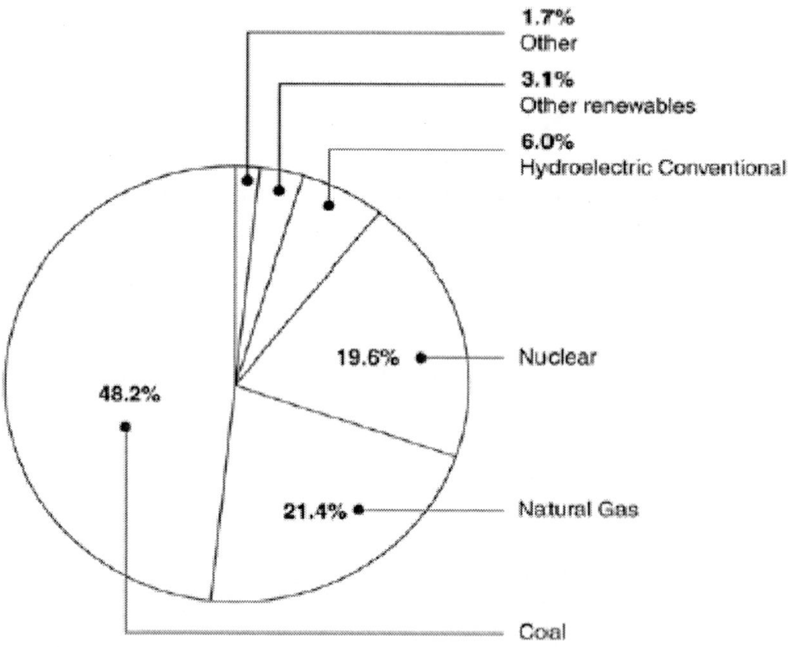

Source: GAO analysis of U.S. Energy Information Administration, Form EIA-923, "Power Plant Operations Report" January 21, 2010.

Figure 1. U.S. Power Generation by Fuel Type, 2008

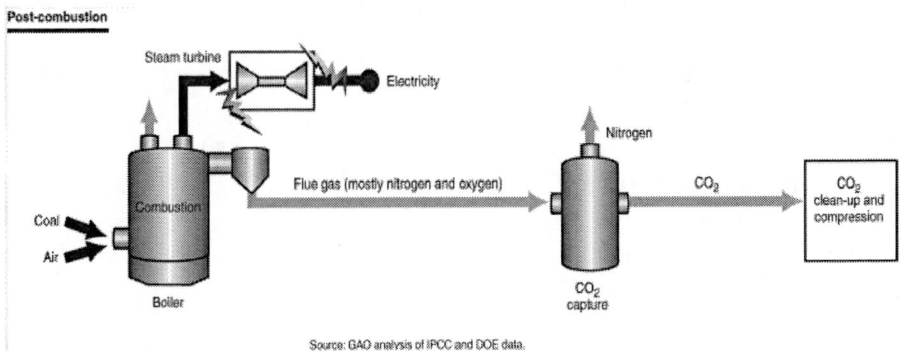

Figure 2. Post-combustion Capture

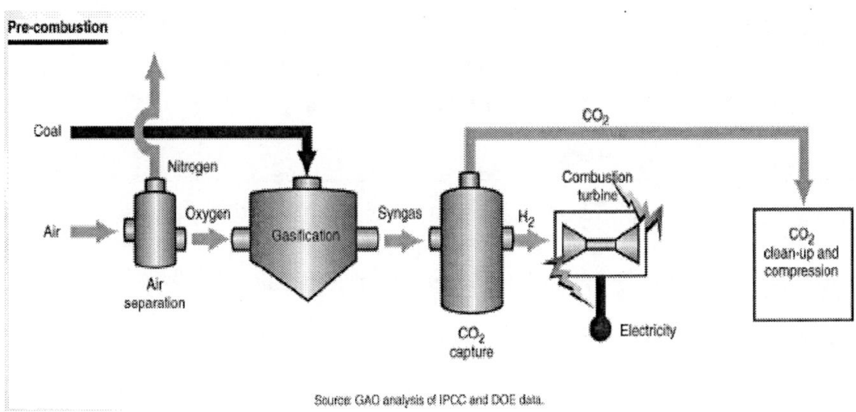

Figure 3. Pre-combustion Capture

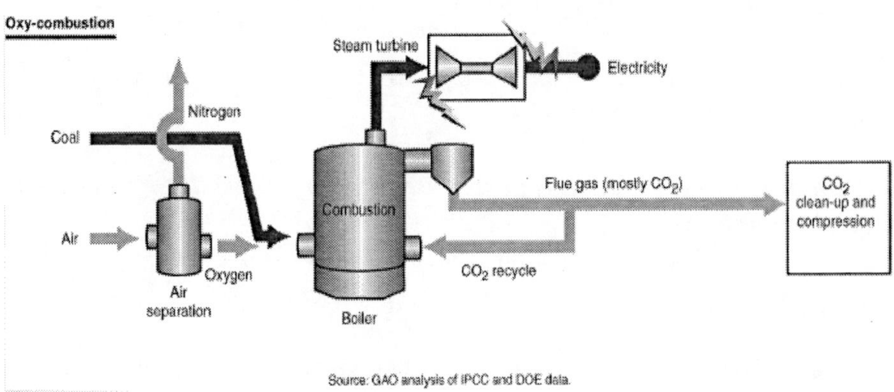

Figure 4. Oxy-combustion Capture

Pre-combustion captures CO_2 produced when gasifying coal

- Compatible with Integrated Gasification Combined Cycle (IGCC) plants, which are in limited use in electricity industry
- The gasification process transforms coal into a syngas, a mixture of hydrogen and carbon monoxide (CO). The CO is then converted into CO_2 and captured

Oxy-combustion captures CO_2 produced when burning coal in oxygen-rich environment

- Compatible with traditional pulverized coal plants, which make up nearly all coal plants currently operating worldwide

The other key technology for reducing CO_2 emissions improves the efficiency of coal plants (efficiency technologies) by allowing plants to use less coal and therefore reduce their CO_2 emissions

- Existing plants
  - Are about 32.5% efficient on average in the U.S. according to a recent Department of Energy (DOE) analysis[27]
  - Can be upgraded to improve efficiencies by a few percentage points
- New plants
  - Can use more efficient designs, such as ultrasupercritical[28]—which operate at higher temperatures and greater steam pressures than conventional plants—and IGCC plants
  - Can achieve efficiencies of 40-44%

CCS and efficiency technologies can be used independently, or in conjunction with one another

Regulatory Efforts and Proposed Legislation Seek to Reduce CO_2 Emissions in U.S.

The Environmental Protection Agency (EPA) is taking steps to regulate CO_2 and other GHGs under the Clean Air Act including

- developing policy guidance to assist permitting agencies in making best available control technology determinations for GHGs that consider the commercial availability of CCS

The American Clean Energy and Security Act, H.R. 2454, passed the House of Representatives on June 26, 2009. Among other things, the proposed legislation would

- establish a GHG cap and trade program
- require new coal power plants permitted before 2020 to reduce CO_2 emissions by half, 4 years after specified CCS deployment criteria are met or 2025, whichever comes first[29]

Federal Investments in Coal Research, Development, and Demonstration (RD&D) Aim to Reduce CO_2 Emissions

DOE's Office of Fossil Energy oversees coal RD&D

- conducts research to accelerate the availability of key coal technologies
- partners with industry and others to move research toward commercialization

Information on Selected DOE Coal Programs

Fossil Energy coal programs	Technology focus
Innovations for Existing Plants	Develop cost-effective post-combustion and oxy-combustion capture technologies
Advanced IGCC	Develop more efficient IGCC plants and integrate these with pre-combustion capture technologies
Carbon Sequestration	Demonstrate storage of CO_2 in geologic formations.[a] Develop improved capture technologies
Advanced Research	Develop technologies to improve plant efficiency, including development of metals for advanced ultrasupercritical plants
Clean Coal Power Initiative (CCPI)	Provide money for commercial demonstration of coal technologies, including CCS

Source: GAO summary of DOE documents.

[a]Geologic formations being examined include saline aquifers, which are composed of porous rock, filled with brine.

In FY09, DOE's coal RD&D funding was at least $681 million

In addition, $3.4 billion was appropriated in the American Recovery and Reinvestment Act for fossil energy RD&D

Objectives

In this context, you asked us to review key technologies to reduce CO_2 emissions from coal power plants

Our objectives were to examine: ·

- the maturity of technologies to reduce CO_2 emissions from coal power plants
- the potential for these technologies to be used commercially in the future and challenges, if any, to their use
- the possible implications of deploying these technologies

Scope and Methodology

Reviewed key reports from

- DOE's national laboratories
- National Academy of Science (NAS)
- Intergovernmental Panel on Climate Change (IPCC)
- International Energy Agency (IEA)
- Global CCS Institute
- National Coal Council
- Academic reports

Conducted scoping interviews with many stakeholders, such as power plant operators, technology vendors, and federal officials

From these scoping interviews, we selected 19 key stakeholders with expertise in coal technologies and asked them a standard set of questions. This group of stakeholders included those from

- Major electric utilities that are planning or implementing projects using key coal technologies
- Technology vendors that are developing these technologies
- Federal officials providing RD&D funding for these technologies
- Researchers from academia and industry that are researching these technologies

Reviewed DOE budget documents and program goals for its RD&D program and interviewed senior DOE staff on these

Visited coal power plants and research facilities in three selected states—AL, MD, and WV[30]

We conducted this performance audit from July 2009 through May 2010, in accordance with generally accepted government auditing standards. Those standards require that we plan and perform the audit to obtain sufficient, appropriate evidence to provide a reasonable basis for our findings and conclusions based on our audit objectives. We believe that the evidence obtained provides a reasonable basis for our findings and conclusions based on our audit objectives.

Results in Brief

DOE does not systematically assess the maturity of key coal technologies, but we found consensus among stakeholders that CCS is less mature than efficiency technologies

Commercial deployment of these technologies is contingent on overcoming economic, technical, and legal challenges

CCS technologies offer more potential to reduce CO_2 emissions than efficiency improvements alone, and both could have cost and other effects

Objective 1. Maturity of Key Technologies

DOE Does Not Systematically Assess their Maturity, but We Found Consensus that CCS Is Less Mature than Efficiency Technologies

DOE does not systematically assess the maturity of key coal technologies, although tools for doing so are available

- DOE does not systematically assess maturity of key coal technologies
- Other agencies charged with developing technologies use Technology Readiness Levels (TRL) to characterize technologies' maturity[31]
- DOE acknowledges TRLs as key practice in contracting, and some DOE offices use this tool for other technology programs

We found consensus among stakeholders that CCS technologies are less mature than efficiency technologies in coal plants

- Key aspects of CCS for use in coal plants still under development
- Efficiency technologies in commercial use

Federal standards for internal control require agency managers to compare actual program performance to planned or expected results and analyze significant differences

DOE's Office of Fossil Energy does not systematically assess the maturity of key coal technologies as they progress toward commercialization

- The agency does not use a standard set of benchmarks or terms to describe the maturity of technologies
- DOE's goals for advancing these technologies sometimes use terms that are not well defined
- DOE officials reported that individual programs are aware of the maturity of technologies, but we found the agency does not formally report on the maturity of these technologies as they progress to commercial scale

Lack of an assessment or benchmarks limits

- DOE's ability to provide a clear picture of the maturity of these technologies to policymakers, utility officials, and others
- Congressional and other oversight of the hundreds of millions DOE is spending on these technologies
- Policymakers' ability to assess the maturity of CCS and the resources that might be needed to achieve commercial deployment

Other Agencies Charged with Developing Technologies use TRLs to Characterize Technologies' Maturity

TRLs were developed by National Aeronautics and Space Administration (NASA), and the agency began using them in the mid 1990s

In 2002, Department of Defense (DOD) specified TRLs as preferred method to conduct technology assessments of weapons programs

TRLs provide a standardized terminology to rank and describe maturity of technologies on a scale of 1 to 9

TRL	Summary of TRL descriptions used by NASA
9	Actual system "flight proven" through successful mission operations under operational mission conditions
8	Actual system completed and "flight qualified" through test and demonstration. Examples include developmental test and evaluation of the system in its intended weapons system to see if it meets design specifications
7	System prototype demonstration in realistic environment. Requires demonstration of actual system prototype in a realistic environment, such as an aircraft vehicle or space
6	System/subsystem model or prototype demonstration in a relevant environment
5	Component and/or breadboard validation in a relevant environment, which could be lab or simulated realistic environment
4	Component and/or breadboard validation in lab environment
3	Proof of concept in lab environment
2	Technology concept and/or application formulated
1	Basic principles observed and reported

Source: GAO analysis of NASA data.

DOE Has Acknowledged TRLs as Key Practice in Contracting, and Some DOE Offices are Using This Tool

DOE's *Technology Readiness Assessment Guide* lays out three key steps in making a technology readiness assessment during the contracting process[32]

- Identify critical technology elements that are essential to successful operation of the facility
- Assess maturity of these critical technologies using TRLs
- Develop a technology maturity plan which identifies activities required to bring technology to desired TRL level
  - Describes current state of technology
  - Describes schedule and budget to move technology to necessary readiness level

Use of the *Guide* is not mandatory

DOE offices have begun to use the *Guide* or TRLs

- Office of Environmental Management uses the *Guide* as part of managing its procurement activities—a result of a GAO recommendation
- Office of Nuclear Energy has begun using TRLs to measure and communicate risks associated with using critical technologies in a novel way

- National Nuclear Security Administration has used TRLs recently as well

We Asked Stakeholders With Expertise in Technologies to Gauge Maturity Using a Scale Based on TRLs

In the absence of a DOE assessment of maturity, we developed a scale for coal technologies based on TRLs in consultation with the Electric Power Research Institute (EPRI), which used a similar approach recently[33]

GAO has used TRLs to gauge the maturity of technologies

TRL	Description of TRLs we developed for coal technologies
9	**Commercial operation** in relevant environment (500 megawatt [MW] coal plant or greater, or about 3 million tons of CO_2captured, transported, or stored annually)[a]
8	**Demonstration** at more than 25% commercial scale (at least 125 MW coal plant, or about 575,000 tons of CO_2 captured, transported, or stored annually)
7	**Pilot scale** at more than about 5% commercial scale (at least 20 MW coal plant, or 100,000 tons of CO_2 captured, transported, or stored annually)
6	**Process development** unit at between about 0.1% to 5% of commercial scale (between 0.5 MW and 20 MW coal plant, or between about 3,000 and 100,000 tons of CO_2 captured, transported, or stored annually)
5	Component validation in relevant environment (coal plant)
4	Component tests in lab
3	Proof of concept test
2	Application formulated (on paper)
1	Basic principles observed

[a] Actual CO_2 emissions can vary based on several factors, including how often a plant operates.

Objective 1. Maturity of CCS Technologies

Components of CCS Widely Used in Other Industries, and a Few Integrated CCS Projects Are Operating

CO_2 capture widely used in natural gas and chemical industries

- CO_2 captured while refining natural gas
- CO_2 captured when gasifying coal to make chemical products such as fertilizer, hydrogen, and synthetic natural gas

CO_2 transported and injected for enhanced oil recovery (EOR) for over 35 years

- CO_2 injected underground to help increase amount of oil recovered
- EOR operations in the U.S. inject about 50 million tons of CO_2 annually, about half of which remains stored underground, according to oil industry officials
- EOR highlighted as a beneficial reuse of captured CO_2

There are a few integrated CCS projects in these industries

- Sleipner and Snohvit (located in North Sea) and In Salah (located in Algeria) are natural gas processing facilities
    - All capture about 1 million tons of CO_2 annually and store it in saline aquifers
- Great Plains Synfuels Plant, located in North Dakota
    - Captures over 3 million tons of CO_2 and transports about 2 million tons of CO_2 annually to the Weyburn oil field in Canada for EOR use

Stakeholders Reported CO_2 Capture at Coal Plants Is at Small Scale

Post-combustion capture

- Stakeholders generally reported largest demonstration is at pilot scale (TRL 7) using our scale
- Largest project taking place is at Mountaineer Plant in WV, which aims to capture over 100,000 tons of CO_2, according to DOE and EPRI

Pre-combustion capture

- Stakeholders offered a range of views on maturity from formulations on paper (TRL 2) to commercial (TRL 9)
    - Some stakeholders said technology is commercial in other industries similar to IGCC plants, such as the Great Plains Synfuels Plant, which captures 3 million tons of CO_2

- Other stakeholders said that pre-combustion capture had not been demonstrated in an IGCC plant and that capturing a large proportion of CO_2 at an IGCC plant required further demonstration of a class of turbines suitable for use with hydrogen fuels

Oxy-combustion capture

- Stakeholders about evenly split between ranking maturity as pilot scale (TRL 7) or process development unit (TRL 6)
- Largest project taking place is at Schwarze Pumpe in Germany, which is a 10 MW scale and aims to capture about 75,000 tons of CO_2 annually according to DOE and EPRI

Stakeholder views on maturity are generally consistent with a 2009 report by the Global CCS Institute that used TRLs[34]

Only One Integrated CCS Project Operating in a Coal Plant, but DOE Has Announced Funding for Additional Projects

The only integrated CCS project in a coal power plant is the Mountaineer Plant in WV according to stakeholders[35]

Project name	DOE award	Project goals
NRG	$154 million	Construct 60 MW demonstration facility using post-combustion capture technology, with captured CO_2 to be used for EOR.
Mountaineer	$334 million	Capture 90% of CO2 from 235 MW flue gas slipstream on 1300 MW plant using post-combustion capture technology. 1.65 million tons of CO_2 captured annually will be stored in nearby saline aquifer.
Texas Clean Energy Project	$350 million	Build 400 MW IGCC plant and capture 90% of CO_2 using pre-combustion capture technology. Over 2.9 million tons of CO_2 captured annually will be used for EOR.
Antelope Valley	$100 million	Capture 90% of $CO_2$from 120 MW flue gas slipstream at existing 450 MW plant. One million tons of CO_2 will be captured annually and could be used for EOR or stored in saline aquifers.
Hydrogen Energy	$308 million	Build advanced IGCC plant that is 250 MW and capture 2 million tons of CO_2 annually to be used for EOR.

Source: GAO summary of CCPI funding announcements.

Note: Additional integrated CCS projects are planned around the world, but not yet operating.

- CO_2 is captured from slipstream of plant's total exhaust with goal of capturing, transporting, and storing over 100,000 tons annually
- Equal to about 20 MW capacity (1.5% of total plant output)

DOE has announced funding for five integrated CCS projects in coal plants through the CCPI (see table below)

CO_2 Compression and Transport Commercially Demonstrated

Nearly all stakeholders reported CO_2 compression and transport demonstrated commercially (TRL 9)

CO_2 is commonly compressed as part of transporting CO_2

There are more than 3,900 miles of pipelines used to transport CO_2 in the U.S.

- These pipelines are primarily used to transport CO_2 for EOR projects in certain areas of the U.S.
- Some of these pipelines have the capacity to transport between 2-10 million tons of CO_2 annually

CO_2 Storage in Oil Reservoirs Considered More Mature than Storage in Saline Aquifers

CO_2 widely injected into oil formations to enhance recovery— resulting in some storage

- Stakeholders generally considered technology commercially demonstrated (TRL 9)
- About 50 million tons of CO_2 injected annually to stimulate additional recovery of oil from wells, and about half remains stored in the formation initially[36]

CO_2 storage in saline aquifers still being demonstrated

- Stakeholders about evenly split between describing maturity at commercial (TRL 9) or demonstration scale (TRL 8)
- Two industrial projects (Sleipner and In Salah) have been injecting over 1 million tons of CO_2 annually into saline formations

- DOE's Sequestration Program has 7 projects that aim to store over 1,000,000 tons of CO_2 in the future, and the majority of these projects are to begin injecting into saline aquifers in 2011 or later

Objective 1. Maturity of Key Technologies

Efficiency Improvements Have Been Deployed Commercially at New and Existing Plants

Efficiency technologies deployed at new plants

- Most stakeholders considered ultrasupercritical and IGCC plants commercially demonstrated (TRL 9)
- A few stakeholders considered IGCC plants less mature than ultrasupercritical
- A number of ultrasupercritical plants ranging from 600 to over 1,000 MW have been built or are under construction in Europe and Asia, with one under construction in U.S.[37]
- Five IGCC plants are operating globally, including two in the U.S. with another under construction[38]

Efficiency technologies deployed at existing plants

- Stakeholders told us efficiency upgrades had been deployed at commercial scale (TRL 9)
- About 10% of U.S. coal plants undertook large efficiency improvements between 1998 and 2008 according to DOE analysis[39]

Objective 2. Challenges to Use of Key Technologies

Commercial Deployment Possible, but Contingent on Overcoming Economic, Technical, and Legal Challenges

Commercial deployment of CCS possible within 10-15 years, but faces major challenges according to reports and stakeholders

- Current CCS technologies are costly to install and operate

- Demonstration of large scale integrated CCS systems needed to assure stakeholders
- U.S. lacks national carbon policy or legal framework to govern CO_2 storage

Many efficiency technologies have been used and are available for commercial use, but still face challenges

- High efficiency coal plants may not be cost-effective
- Some higher efficiency plant designs not fully demonstrated and require advanced materials
- Improvements made to existing plants may trigger additional regulatory requirements

Objective 2. Challenges to Use of Key Technologies

Several Groups Expect CCS Deployment in 10-15 Years

Stakeholder group	Goals for commercial deployment of CCS
DOE	Widespread, affordable deployment of CCS should begin in 8-10 years
National Coal Council	By 2020, deployment of CCS in 5 to 7 gigawatts worth of power plants as part of a "pioneer phase of deployment"
IEA	Commercial deployment of CCS should begin by 2025
Coal Utilization Research Council, an industry group, and EPRI	Identifies family of technologies to reduce emissions from coal plants, including CCS and efficiency technologies by 2025

Source: GAO summary of relevant reports.

Objective 2. Challenges to Use of CCS Technologies

Economic: Current CCS Technologies are Costly to Install and Operate

Current CCS technologies are costly to install

- In 2007 DOE estimated initial capital investment costs could be
  - 85% higher for plants with post-combustion capture and
  - 36% higher for pre-combustion capture at IGCC plants, compared to comparable plants without CCS[40]
- Electric utilities not likely to adopt costly technologies without assured cost recovery

Current CCS technologies require significant energy to operate, reducing the electricity plants can sell and raising operating costs

- Parasitic loads—energy used onsite—for current CCS technologies are estimated to be
  - between about 21% and 32% of plant output for post-combustion
  - between 15% and 22% of plant output for pre-combustion[41]
- DOE devoting R&D money to develop novel CO_2 capture technologies to lower the parasitic load, but these remain at smaller scale
  - DOE is funding post-combustion work on membranes, sorbents, and solvents in the hope of lowering the current cost of CO_2 capture
  - Research is also being conducted on using captured CO_2 to grow algae, a potential liquid transportation fuel

Technical: Demonstration of Large Scale Integrated CCS Systems Needed to Assure Stakeholders

Key studies report that demonstration of large scale integrated CCS systems is needed to

- Demonstrate the performance and potential costs of these systems
- Gain experience in designing and building systems to help drive down the costs of these technologies

Some stakeholders also reported that additional demonstration needed to lower perceived risk of technologies

- Officials from one large electric utility told us that demonstration projects were needed to build experience with the technologies and to build vendor confidence so that they could provide technology performance guarantees
- Officials from one state public utility commission reported that they considered CCS immature and were unlikely to approve cost recovery for a plant with CCS in the foreseeable future
- Officials from two financial firms reported that they considered the application of CCS technologies at coal plants largely unproven and they would require additional demonstration projects and performance guarantees from technology vendors to help reduce the risk of financing these projects

Legal: U.S. Lacks National Carbon Policy or Legal or Regulatory Framework to Govern CO_2 Storage

Without national carbon policy, nearly all stakeholders said CCS would not be widely deployed

- Without a tax or a sufficiently restrictive limit on CO_2 emissions, plant operators lack economic incentive to reduce emissions
- Reports by IPCC, NAS, and Global CCS Institute have all highlighted the importance of a carbon policy to incentivize the use CCS
- Such a policy driver could help to accelerate the development of CCS

Lack of a regulatory framework for storing CO_2 and uncertainty regarding liability for stored CO_2 are also challenges

- Nearly all stakeholders reported these as large or very large challenges to storing CO_2
- EPA to issue a final rule for injection of CO_2 for geologic sequestration in fall 2010 under the Safe Drinking Water Act (SDWA)
    - EPA lacks authority to release well operators from liability for endangerment of underground sources of drinking water until the operator meets all of the closure and post-closure requirements and EPA approves site closure

    - Once site closure is approved, operators are only liable under the SDWA for violating or failing to comply with EPA orders in situations that pose an imminent and substantial endangerment
    - Potential storage site operators are unlikely to assume this risk
- EPA's rule will not address who is liable for unintended releases of stored CO_2 that have other harmful effects
- Determining ownership of subsurface pore space presents additional challenge

Objective 2. Challenges to Use of Efficiency Technologies

Economic: High Efficiency Coal Plants May Not Be Cost-Effective

Low prices for coal and other fuels in the U.S. may limit the incentive to build more efficient, but costly, plants

- Ultrasupercritical plants have higher capital costs because they use advanced materials, which may not justify expected fuel savings
- IGCC plants are more expensive than pulverized coal units, and there are few in operation globally
- If low natural gas prices persist, utilities may choose to build natural gas power plants to reduce CO_2 emissions in lieu of efficient coal plants

Incentives complicate construction of more efficient plants in regulated states

- Building new, more efficient coal plants faces hurdles
    - State utility commission approval required to build new plants
    - Demonstrating merits of more efficient plants may be difficult
- Fuel clauses may limit utility interest in fuel savings
    - Some utilities can "pass through" coal price increases to customers using fuel adjustment clauses

To date, all of the more efficient ultrasupercritical plants have been built outside the U.S., where coal prices are generally higher

A tax or limit on CO_2 emissions could increase the price of coal and help to incentivize the adoption of efficiency technologies

Technical: Some Higher Efficiency Plant Designs Not Fully Demonstrated and Require Advanced Materials

Some advanced power plant designs require materials that can withstand more extreme conditions than those found in current plants

"Advanced" ultrasupercritical plants require development of metal alloys to withstand steam temperatures that could be 300 to 500 degrees Fahrenheit higher than today's ultrasupercritical plants[42]

Advanced IGCC plants require development of certain components, including more efficient ways to generate oxygen and improved gasifiers that can gasify coal at higher pressures

Regulatory: Improvements May Subject Existing Plants to Additional Regulations

Most stakeholders said the Clean Air Act's New Source Review (NSR) requirements limit efficiency improvements at existing plants

- NSR is triggered when a company constructs new facilities or makes a major modification—a physical or operational change that would result in a significant net increase in emissions
- Under NSR, permitting authorities establish emissions limits for the facility and ensure the appropriate pollution controls will be used

Several stakeholders said that utilities could improve their plants' efficiency but were reluctant to do so because they feared this would trigger NSR which could require the installation of costly pollution controls

Objective 3. Implications of Using Key Technologies

CCS Offers More Potential to Reduce CO_2 Emissions than Efficiency Improvements Alone, and Both Could Have Cost and Other Effects

CCS has positive and negative implications

- A key advantage is that CCS could help meet GHG limits and allow coal to remain part of the nation's fuel mix
- The use of CCS raises some key concerns
  - Electricity costs and demand for water could increase[43]
  - Could affect ability of individual plants to operate reliably

Technologies to improve the efficiency of coal plants have positive and negative implications

- A key advantage is that plant efficiency improvements offer more potential for near term emissions reductions
- The use of efficiency technologies raises some concerns
  - Unlikely to meet ambitious cuts in CO_2 by themselves
  - Stakeholders had mixed views on other potential effects, such as cost

Objective 3. Implications of Using CCS

CCS Could Help Meet GHG Limits and Allow Coal to Remain Part of Fuel Mix

Key reports have highlighted the key role that CCS could have in meeting potential limits on GHG emissions

- EPRI – Estimated that CCS could help meet 12% of reductions needed to reduce emissions in electricity sector by 41% by 2030[44]
- IEA – Estimated that CCS could meet 20% of reductions needed to reduce global CO_2 emissions by half by 2050[45]
  - Both studies note that cost of meeting these limits would increase if CCS not deployed

CCS could allow coal to remain part of fuel mix according to stakeholders and reports

- Majority of stakeholders said CCS would allow coal to remain part of fuel mix for generating electricity
- Massachusetts Institute of Technology (MIT) researchers called CCS the "critical enabling technology" to reduce CO_2 emissions while allowing continued use of coal in the future[46]
- NAS stated if CCS does not develop, electricity sector could move more towards using natural gas to meet emissions targets[47]
- GAO's past work found that switching from coal to natural gas could lead to higherfuel costs, and increased exposure to the greater price volatility of natural gas[48]

Objective 3. Implications of Using CCS

CCS Could Increase Electricity Costs and Water Demand

Most stakeholders told us that CCS would likely increase electricity costs. In addition, key reports have estimated potential cost increases

- MIT estimated that plants with post-combustion capture have 61% higher cost of electricity, and IGCC plants with pre-combustion capture have a 27% higher cost[49]
- DOE estimated that plants with post-combustion capture have 83% higher cost of electricity, while IGCC plants with pre-combustion capture have a 36% higher cost[50]

DOE has raised concerns about water consumption associated with CCS

- DOE estimated that post-combustion capture technology could almost double water consumption at a coal plant, while pre-combustion capture could increase water use by 73%
- DOE officials said that continued development of CCS and cooling technologies could significantly reduce water use for CCS

CCS Could Compromise Reliability

Some utility officials said CCS could lead to decline in reliability of individual plants

- A power plant might need to shut down if any of the three components (capture, transportation, storage) of CCS became unavailable
- Such unplanned shutdowns could impact reliability of electric supply

Other sources of electricity would need to make up for the parasitic load associated with CCS

National Coal Council reported temporary declines in reliability during past deployments of new coal technologies[51]

Objective 3. Implications of Using Efficiency Technologies

Plant Efficiency Improvements Offer Potential for Near Term Emissions Reductions but Raise Some Concerns

Plant efficiency improvements offer potential for near term emissions reductions

- Making efficiency upgrades to existing fleet can happen much sooner than building new, more efficient plants
- DOE estimates that efficiency improvements could reduce CO_2 emissions by 100 million tons annually, about an overall 5-10% reduction in fleet emissions
- According to National Coal Council, increasing efficiency is "only practical method for mitigating CO_2 emissions now" in coal plants[52]

Plant efficiency improvements alone cannot reduce CO_2 emissions from a coal plant to the same extent as CCS according to DOE and others

- Ultrasupercritical coal plant with 44% efficiency will emit about a one-third less CO_2 than an average U.S. plant
- Upgrades made to existing plants can improve efficiency by a few percentage points, resulting in a decline in CO_2 emissions from the plant by about 5-10%
- CCS offers potential to capture 90% of a plant's CO_2 emissions
- Efficiency improvements can, however, facilitate CCS because they help reduce the amount of CO_2 to be handled by the CCS system

Stakeholders had mixed views on other potential effects

- Stakeholders' views were mixed on potential effect on electricity costs
- Stakeholders generally did not think efficiency technologies would increase water demand or compromise reliability

Concluding Observations

Addressing climate change while retaining the use of coal power plants will likely require the successful deployment of new technologies

- CCS, in particular, remains relatively immature compared to efficiency technologies
- Some of the discussions surrounding regulatory efforts and proposed climate change legislation have focused on the commercial availability of CCS technologies
- DOE plays a key role in helping to accelerate commercial availability of CCS technologies and is spending hundreds of millions of dollars annually for this effort
- Standards for internal controls require agency managers to compare actual program performance to planned or expected results and analyze significant differences
- DOE is not systematically assessing the maturity or progress of CCS or other advanced coal technologies toward commercialization
- As a result, DOE cannot provide
  - A clear picture of the maturity of technologies, and resources needed to achieve commercial demonstration
  - Critical information for policymakers as they consider climate change policies

Potential Next Steps for DOE

Develop a standard set of benchmarks to gauge the maturity of key coal technologies and report to Congress on the maturity of these technologies

Consider using its *Technology Readiness Assessment Guide* to develop benchmarks and reporting requirements for coal technologies

Appendix II. Scope and Methodology

To conduct this work, we reviewed key reports including those from the Department of Energy's (DOE) national laboratories, the National Academy of Sciences, International Energy Agency (IEA), Intergovernmental Panel on Climate Change, Global CCS Institute, the National Coal Council, and academic reports.

To identify stakeholders' views on these technologies, we conducted initial scoping interviews with power plant operators, technology vendors, and federal officials from the Environmental Protection Agency (EPA) and DOE.

Following this initial round of interviews, we selected a group of 19 stakeholders with expertise in carbon capture and storage (CCS) or technologies to improve coal plant efficiency and asked them a set of standard questions. This group of stakeholders included representatives from major utilities that are planning or implementing projects that use these technologies, technology vendors that are developing these technologies, federal officials that are providing research, development, and demonstration funding for these technologies, and researchers from academia or industry that are actively researching these technologies.

During these interviews, we asked stakeholders to describe the maturity of technologies in terms of a scale we developed, based on Technology Readiness Levels (TRL). TRLs are a tool developed by the National Aeronautics and Space Administration and used by various federal agencies to rate the extent to which technologies have been demonstrated to work as intended using a scale of 1 to 9. In developing TRLs for coal technologies, we consulted with the Electric Power Research Institute (EPRI), which had recently used a similar approach to examine the maturity of coal technologies.[53] Specifically, EPRI developed specific benchmarks to describe TRLs in the context of a commercial scale coal power plant. For example, they defined TRL 8 as demonstration at more than 25 percent the size of a commercial scale plant. We applied these benchmarks to a commercial scale power plant, which we defined as 500 megawatts (MW) and emitting about 3 millions tons of carbon dioxide (CO_2) annually. We based this definition on some of the key reports we reviewed, which used 500 MW as a standard power plant, and stated that such a plant would emit about 3 million tons of CO_2. Actual CO_2 emissions from a power plant can vary based on a variety of factors, including the amount of time that a power plant is operated. We also reviewed available data on the use of key coal technologies compiled by IEA and the Global CCS Institute.

To identify the potential for these technologies to be used commercially in the future along with any associated challenges or implications, we reviewed key reports on CCS and efficiency technologies. We also examined reports developed by DOE, IEA, and electricity industry groups, which lay out goals for the deployment of advanced coal technologies to reduce CO_2 emissions. We also used our interviews with stakeholders with expertise on these technologies to seek their views on the potential challenges to the commercial deployment of these technologies and implications that could be associated with their use.

Finally, we conducted site visits to coal power plants and research facilities in three states—Alabama, Maryland, and West Virginia. We selected this nonprobability sample of states because they contained projects involving advanced coal technologies. During these visits, we interviewed utilities and technology vendors about the goals for these projects along with any challenges they were encountering.

APPENDIX III. COMMENTS FROM THE DEPARTMENT OF ENERGY

Note: GAO comments supplementing those in the report text appear at the end of this appendix. Page numbers in draft report may differ from those in this report.

Department of Energy
Washington, DC 20585

June 4, 2010

Mr. Mark E. Gaffigan
Director
Natural Resources and Environment Team
U.S. Government Accountability Office
441 G Street, NW, Mail 2T23A
Washington, DC 20548

Dear Mr. Gaffigan:

Thank you for the opportunity to review the Government Accountability Office (GAO) draft report entitled, "Coal Power Plants: Opportunities Exist for DOE to Provide Better Information on the Maturity of Key Technologies to Reduce Carbon Dioxide Emissions" (GAO-10-675). Enclosed pleased find the U.S. Department of Energy's comments on the draft report.

If you have any questions or comments please contact Dr. Darren Mollot of my staff at (301) 903-2700.

Sincerely,

James J. Markowsky
Assistant Secretary
Office of Fossil Energy

Enclosure:
DOE Comments on Draft GAO Report

See comment 1.

Department of Energy Comments on GAO "Coal Power Plants: Opportunities Exist for DOE to Provide Better Information on the Maturity of Key Technologies to Reduce Carbon Dioxide Emissions (GAO-10-675) (GAO Draft Report)

This responds to your request for comments by the Department of Energy on the above referenced GAO Draft Report.

Our main response is that we agree with GAO's statement that DOE plays a "key role" in working to advance carbon capture and storage (CCS) and efficiency technologies toward commercialization and in helping policy makers have an accurate view of their maturity (pages 3, 14, 53). However, we take some exception with GAO's assessment that DOE is unable to provide a clear picture of the maturity of these technologies or quantify the necessary resources that might be required to move these technologies toward commercial demonstration (this assertion can be found in several places within the report including pages 6, 14, 30, 53).

The Office of Fossil Energy (FE) acknowledged that it could improve upon its current process of providing a clearer picture of technology maturity. FE, working with several national labs, conducts a great deal of research, development, and demonstration activities on CCS and other carbon reduction technologies. These efforts are reported, analyzed, reviewed, and studied, the outcomes from which frequent and continued assessments are made regarding the current status of CCS technologies including their commercial readiness. Included in many of these assessments are specific development timelines that can be used to ascertain how long development will take, and how much development will likely cost. These activities also feed into the development of CCS Roadmaps (e.g. Carbon Sequestration Technology Roadmap and Program Plan http://www.netl.doe.gov/technologies/carbon_seq/refshelf/project%20portfolio/2007/2007Roadmap.pdf,) for the specific development of core CO_2 reduction technologies. These activities and others collectively paint a very accurate picture of the current status of CCS technologies, and provide an excellent gauge on what resources would need to be committed in order to achieve deployment of CCS by the 2020 timeframe.

Furthermore, GAO made the following recommendation beginning on page 14 (also repeated on pages 6-8, and the lower left hand corner of the front cover):

> ***"We recommend that the Secretary of Energy direct the Office of Fossil Energy to develop a standard set of benchmarks to gauge the maturity of key technologies and report to Congress on the maturity of these technologies. As part of this process, the Office of Fossil Energy should consider consulting DOE's Technology Readiness Assessment Guide to develop benchmarks and reporting requirements."***

Consistent with our continued efforts to supply policy makers with clear information in a form more amenable for them to gauge the maturity of CCS technologies, we concur

See comment 2.

See comment 3.

See comment 4.

Corrective Action Plan, and will, at regular intervals, report to Congress on the status of its actions toward instituting GAO's recommendation.

Specific clarifying comments on the GAO report are as follows:

1 On page 28, the report states:

> ***"DOE does not systematically assess the maturity of key coal technologies, but we found consensus among stakeholders that CCS is less mature than efficiency technologies"***

Although DOE has not assessed the maturity of coal technologies using technology readiness levels (TRLs), we are very aware of the maturity of all the technologies in the portfolio. We plan to do a formal TRL assessment in the near future.

It would also be important to clarify the precise meaning of the term, "efficiency technologies" in this context. There is a considerable difference between improving coal plant efficiencies as opposed to more efficient light bulbs

Furthermore, this is an overly broad statement. For example, one could argue that post-combustion CO_2 capture via amine scrubbing, plus enhanced oil recovery (EOR) for CO_2 storage, is much more mature than 1400°F ultrasupercritical "efficiency" technology

2. In response to the comments on page 30, we would like to point out that we are currently in compliance with the policy stated in bullet point #1:

> ***"Federal standards for internal control require agency managers to compare actual program performance to planned or expected results and analyze significant differences"***

The way this particular set of points is organized gives the appearance that DOE may not be fulfilling this Federal standard. However, variances are routinely calculated and analyzed in accordance with these standards.

3. On page 35, the report states that there is:

> ***"Only One Integrated CCS Project Operating in a Coal Plant"*** and goes on to state, ***"The only integrated CCS project in a coal plant is the Mountaineer Plant in WV according to stakeholders"***

There is a second integrated CCS project operating in a coal plant It is the Great Plains Gasification plant. In this project, the CO_2 is captured and piped to Canada for

See comment 5.
See comment 6.
See comment 6.

4. Page 39 discusses several issues related to efficiency technologies being deployed at new plants:

> ***"Most stakeholders considered ultrasupercritical and IGCC plants commercially demonstrated (TRL 9)"***
>
> and
>
> ***"A few stakeholders considered IGCC plants less mature than ultrasupercritical"***

With respect to both statements, it is necessary to define the Ultrasupercritical (USC) temperatures being discussed here. The Ultrasupercritical term has been used for plants ranging from 1112°F to 1400°F In addition, the Tampa and Wabash IGCCs are operating commercially today as are other IGCC plants outside the US at TRL=9.

5. The information presented on page 40 is potentially misleading as it attempts to compare the commercial deployability of CCS versus efficiency technologies. Efficiency improvements may give a few percentage points of improvement resulting in perhaps 10% reduction in CO_2 emissions, whereas commercial deployment of CCS could result in 90% or greater CO_2 reductions. Also, CCS should be deployable by 2020. Moreover, CCS can be considered to be essentially deployable today with IGCC plus EOR, albeit with additional integration risks and financial incentive to do so. In summary, while the information on page 40 is accurate from a certain point of view, the perspective infers that the two approaches can be easily compared whereas the scope and benefits of the two approaches are very different.

6. Page 48 summarizes the positive and negative implications of using CCS versus efficiency technologies With respect to the advantages of using plant efficiency improvements, the following statement is made:

> ***"A key advantage is that plant efficiency improvements offer more potential for near term emissions reductions"***

While this statement may be technically true, it is important to note that there is far less potential to achieve the deep CO_2 reductions that will be needed to meet our nation's climate-related goals solely using efficiency technologies. It is important to have that in mind when making these kinds of general statements.

7. On page 50, the report states:

See comment 7.
See comment 8.
See comment 8.

"DOE estimated that post-combustion capture technology could almost double water consumption at a coal plant, while pre-combustion capture would increase water use by 37%"

While this is true, it should also be mentioned that the continued development of advanced CCS and cooling technologies could significantly reduce water use for CCS

8. While the information presented on page 22 is factually correct concerning existing coal plants, it is important to note that in 2008, the U.S. coal-fired power plant (CFPP) fleet had a generation-weighted average efficiency of 32.5% while the top ten percent of the fleet had an efficiency of 37.6%, five percentage points higher. (http://www.netl.doe.gov/energy-analyses/pubs/ImpCFPPGHGRdctns_0410.pdf). Furthermore, "Ultra-supercritical steam parameters of 4350 psi and 1112°F (300 bar and 600°C) are in operation today with generating efficiencies of 40% (HHV). There are several years of experience with these plants in Europe and Japan, with excellent availability, and plans have been announced for several USC PC plants in the United States." http://mydocs.epri com/docs/public/000000000001016877.pdf [see pages 3-5]

Future plants could go much higher in efficiency using USC with 1400°F steam temps or IGCC with solid oxide fuel cells.9. With respect to the assertion on page 46 that advanced ultrasupercritical plants requiring metal alloys that withstand 27% higher steam temperatures, we recommend that absolute temperatures are used. A percent increase in temperature value is only meaningful if you use absolute temperatures (Rankine or Kelvin) To advance from 1116°F to 1300°F using absolute temperatures would be a 12% increase in temperature and going further to 1400°F would be an 18% increase. Under the circumstances, it might be better to just say "a temperature increase of 300°F to 400°F."

The following are GAO's comments on the Department of Energy's letter dated June 4, 2010.

GAO Comments

1. We acknowledge that DOE publishes reports that assess the technical and economic feasibility of some advanced coal technologies and revised our report accordingly. While some of these reports provide valuable information, we found that the agency does not systematically review these technologies, have a standard set of benchmarks to describe the maturity of technologies as they progress to commercialization, or prepare a formal report on a regular basis to assess their maturity or the resources needed to advance technologies toward commercialization. We are encouraged that DOE acknowledges that improvements can be made to the information it provides to policymakers and concurs with our recommendation that the agency develop a standard set of benchmarks and report on the maturity of these technologies to Congress. Finally, the agency notes that it plans to do a formal assessment using TRLs of coal technologies in the near future in line with our recommendation.
2. Our draft report defines efficiency technologies as referring to new power plant designs such as Integrated Gasification Combined Cycle and ultrasupercritical along with efficiency upgrades made to existing coal power plants. The statement in our draft report that CCS is less mature than efficiency technologies in coal power plants is based on stakeholder views of coal technologies using our TRL scale. Our draft report notes that certain aspects of CCS have been used commercially in other industries such as natural gas processing or enhanced oil recovery. In addition, the draft report indicates that one of the challenges to using advanced ultrasupercritical plants is the lack of metal alloys to withstand increased steam temperatures.
3. We are not suggesting that DOE is not complying with this standard. This standard outlines the broad duties federal agencies have in managing their programs. Our finding discussed in our comment one above identifies that DOE could do more to improve its efforts to address this standard.
4. We have revised our draft report to indicate that there is only one integrated CCS project in a coal power plant. DOE states that the Great Plains Synfuels plant is an integrated CCS project. We agree that this plant is capturing and transporting CO_2 to be used as part of enhanced oil recovery in Canada's Weyburn oil field. However, this plant gasifies coal in order to make synthetic natural gas; it is not a

coal power plant that produces electricity, which is the focus of our report.

5. Our report defines ultrasupercritical plants as having steam temperatures of about 1,100 degrees Fahrenheit.
6. We agree with DOE that there is a difference in the ability for CCS and efficiency technologies to achieve reductions in CO_2 emissions from coal power plants. Specifically, our report states that the use of efficiency technologies by themselves are "unlikely to meet ambitious cuts in CO_2." In addition, we state that efficiency technologies cannot reduce CO_2 emissions from the same extent as CCS. For example, we state that an ultrasupercritical plant emits about one-third less CO_2 than an average coal power plant in the United States, while CCS offers the potential to capture 90 percent of a plant's CO_2 emissions.
7. We revised our draft report to note that advancements in CCS and cooling technologies could help to reduce water use for CCS. In addition, it is important to note that our report states that pre-combustion capture could increase water use by 73 percent, not 37 percent as DOE's comment indicates.
8. We have made these technical changes to our draft report. It is important to note that we state that advanced materials are needed to withstand temperature increases of 300 to 500 degrees Fahrenheit. This is because today's ultrasupercritical plants have steam temperatures of about 1,100 degrees Fahrenheit, while DOE has set goals to develop materials to withstand steam temperatures of 1,400 to 1,600 degrees Fahrenheit.

End Notes

[1] EIA is the statistical and analytical agency within DOE that collects, analyzes, and disseminates independent and impartial energy information.

[2] For the purposes of this report, we have defined ultrasupercritical to mean steam temperatures of about 1,100 degrees Fahrenheit.

[3] H.R. 2454, § 311, 111th Cong. (2009).

[4] EPA must determine whether certain CCS deployment criteria are met, including whether commercial power plants and other stationary sources have captured and stored at least 12 million tons of CO_2 annually, to trigger the emission reduction requirement before 2025.

[5] Pub.L. No. 111-5 (2009). One of the stated purposes of the ARRA is to preserve and create jobs and promote economic recovery.

[6] EPRI is an independent nonprofit company funded by electricity producers that conducts research and development in the electricity sector. EPRI's work contributed to the following

report: Global CCS Institute, *Strategic Analysis of the Global Status of Carbon Capture and Storage: Synthesis Report* (Canberra, Australia, 2009).

[7] GAO, *Standards for Internal Control in the Federal Government*, GAO/AIMD-00-21.3.1 (Washington, D.C.: November 1999).

[8] TRLs were developed by NASA and the agency began using them in the mid-1990s. In 2002, DOD specified TRLs as the preferred method to conduct technology assessments for weapons programs.

[9] DOE, *Technology Readiness Assessment Guide*, DOE G413.3-4 (Washington, D.C., Oct. 12, 2009).

[10] GAO, *Department of Energy: Major Construction Projects Need a Consistent Approach for Assessing Technology Readiness to Help Avoid Cost Increases and Delays*, GAO-07-336 (Washington, D.C.: Mar. 27, 2007).

[11] While gasifying coal to make synthetic natural gas, the Great Plains Synfuels plant captures and transports CO_2 for EOR use. However, this plant does not produce electricity.

[12] There is an ultrasupercritical plant under construction in the United States known as the John W. Turk, Jr. plant. This 600 MW plant is being built in Arkansas and is scheduled to be completed in 2012. In addition, there is also a 630 MW IGCC plant under construction in Indiana, known as the Edwardsport plant. This plant is scheduled to be completed in 2012.

[13] Our past work has also highlighted some of the challenges to deploying CCS. See GAO, *Climate Change: Federal Actions Will Greatly Affect the Viability of Carbon Capture and Storage As a Key Mitigation Option*, GAO-08-1080 (Washington, D.C.: Sept. 30, 2008).

[14] DOE, *Cost and Performance Baseline for Fossil Energy Plants–Volume 1: Bituminous Coal and Natural Gas to Electricity, Final Report* (2007).

[15] Under the Underground Injection Control program, EPA regulates underground injections of various substances into injection wells. Currently, CO_2 injection wells can be permitted as Class I (injections of hazardous wastes, industrial nonhazardous wastes, municipal wastewater) or Class V wells (injections not included in other classes, including wells used in experimental technologies such as pilot CO_2 storage). EPA's rule will establish a Class VI well for injection of CO_2 for geologic sequestration.

[16] 42 U.S.C. § 300i.

[17] Today's ultrasupercritical plants have steam temperatures of about 1,100 degrees Fahrenheit. DOE has a goal to develop materials to withstand steam temperatures of 1,400 to 1,600 degrees Fahrenheit.

[18] IEA, *Technology Roadmap: Carbon capture and storage*(Paris, France, 2009).

[19] MIT, *The Future of Coal* (Cambridge, Mass., 2007).

[20] NAS, *America's Energy Future* (Washington, D.C., 2009).

[21] GAO, *Economic and Other Implications of Switching from Coal to Natural Gas at the Capitol Power Plant and at Electricity-Generating Units Nationwide,* GAO-08-601R (Washington, D.C.: June 5, 2006).

[22] MIT, *The Future of Coal.*

[23] DOE, *Cost and Performance Baseline for Fossil Energy Plants–Volume 1.*

[24] DOE, *Cost and Performance Baseline for Fossil Energy Plants–Volume 1.* DOE officials also said that continued development of CCS and cooling technologies could significantly reduce water use for CCS.

[25] National Coal Council, *Low-Carbon Coal: Meeting U.S. Energy, Employment and CO_2 Emission Goals with 21st Century Technologies* (Washington, D.C., December 2009).

[26] National Coal Council Issue Paper, *Higher Efficiency Power Generation Reduces Emissions* (2009).

[27] This analysis also found that the top 10% of the U.S. coal fleet had an average efficiency of 37.6%. See DOE, *Improving the Efficiency of Coal-Fired Power Plants for Near Term Greenhouse Gas Emissions Reductions* (April 16, 2010).

[28] For the purposes of this report, we have defined ultrasupercritical to mean steam temperatures of about 1,100 degrees Fahrenheit.

[29] EPA must determine whether certain CCS deployment criteria are met, including whether commercial power plants and other stationary sources have captured and sequestered at least 12 million tons of CO_2 annually to trigger the emission reduction requirement before 2025.

[30] We selected this nonprobability sample of states because they contained projects involving advanced coal technologies.

[31] TRLs are used to gauge technology maturity and use a 9 point scale to rate the extent to which technologies have been demonstrated to work as intended.

[32] DOE, *Technology Readiness Assessment Guide*, DOE G413.3-4, (Washington D.C., Oct. 12, 2009).

[33] EPRI is an independent nonprofit company funded by electricity producers that conducts research and development in the electricity sector. EPRI's work was part of the following report: Global CCS Institute, *Strategic Analysis of the Global Status of Carbon Capture and Storage: Synthesis Report* (Canberra, Australia, 2009).

[34] Global CCS Institute, *Strategic Analysis of the Global Status of Carbon Capture and Storage: Synthesis Report.*

[35] While gasifying coal to make synthetic natural gas, the Great Plains Synfuelsplant captures and transports CO2for EOR use. However, this plant does not produce electricity.

[36] According to oil industry officials, the other half of the CO_2 is captured during the process of recovering oil to be injected again for EOR. They also reported that the intention of EOR is to recover additional oil, not to store CO_2, but this is an unintended consequence of injecting the CO_2. The Global CCS Institute has reported that experiences with EOR have yielded experience with transporting and injecting CO_2, but have yielded little information on CO_2 storage and long-term monitoring of the stored CO_2.

[37] This ultrasupercritical plant is known as the John W. Turk, Jr. Plant. This 600 MW plant is being built in Arkansas and is scheduled to be completed in 2012.

[38] This IGCC plant is known as the Edwardsport plant. This 630 MW plant is being built in Indiana and is scheduled to be completed in 2012.

[39] DOE, *Improving Efficiency of Coal-Fired Power Plants for Near Term Greenhouse Gas Emissions Reductions* (Feb. 25, 2010).

[40] DOE, *Cost and Performance Baseline for Fossil Energy Plants–Volume 1: Bituminous Coal and Natural Gas to Electricity, Final Report* (2007).

[41] MIT, *The Future of Coal* (Cambridge, Mass., 2007). DOE, *Cost and Performance Baseline for Fossil Energy Plants–Volume 1.*

[42] Today's ultrasupercritical plants have steam temperatures of about 1,100 degrees Fahrenheit. DOE has a goal to develop materials to withstand steam temperatures of 1,400 to 1,600 degrees Fahrenheit.

[43] Water is needed to generate electricity and process fuels to generate electricity. Due to the parasitic load associated with current CCS technologies, more electricity must be produced to supply the same amount of electricity to consumers, leading to additional water consumption. See GAO, *Energy- Water Nexus: Improvements to Federal Water Use Data Would Increase Understanding of Trends in Power Plant Water Use*, GAO-1 0-23 (Washington, D.C.: Oct. 16, 2009).

[44] EPRI, *PRISM/MERGE Analysis* (Palo Alto, California, 2009).

[45] IEA, *Technology Roadmap: Carbon capture and storage*(Paris, France, 2009.)

[46]MIT, *The Future of Coal.*

[47] NAS, *America's Energy Future* (Washington, D.C., 2009).

[48] GAO, *Economic and Other Implications of Switching from Coal to Natural Gas at the Capital Power Plant and at Electricity-Generating Units Nationwide*, GAO-08-601R (Washington, D.C.: May 1, 2008).

[49] MIT, *The Future of Coal.*

[50] DOE, *Cost and Performance Baseline for Fossil Energy Plants–Volume 1.*

[51] National Coal Council, *Low-Carbon Coal: Meeting U.S. Energy, Employment and CO2 Emission Goals with 21st Century Technologies* (Washington, D.C., December 2009).

[52] National Coal Council Issue Paper, *Higher Efficiency Power Generation Reduces Emissions* (2009).

[53] EPRI is an independent nonprofit company funded by electricity producers that conducts research and development in the electricity sector. EPRI's work was part of the following report: Global CCS Institute, *Strategic Analysis of the Global Status of Carbon Capture and Storage: Synthesis Report* (Canberra, Australia, 2009).

In: Coal-Fired Power Plants and Carbon… ISBN: 978-1-61122-054-4
Editor: Matthew W. Shuester

Chapter 3

CLEAN COAL: DOE'S DECISION TO RESTRUCTURE FUTUREGEN SHOULD BE BASED ON A COMPREHENSIVE ANALYSIS OF COSTS, BENEFITS, AND RISKS

United States Government Accountability Office

WHY GAO DID THIS STUDY

Coal-fired power plants generate about one-half of the nation's electricity and about one-third of its carbon dioxide (CO_2) emissions, which contribute to climate change. In 2003, the Department of Energy (DOE) initiated FutureGen—a commercial-scale, coal-fired power plant to incorporate integrated gasification combined cycle (IGCC), an advanced generating technology, with carbon capture and storage (CCS). The plant was to capture and store underground about 90 percent of its CO_2 emissions. DOE's cost share was 74 percent, and industry partners agreed to fund the rest. Concerned about escalating costs, DOE restructured FutureGen. GAO was asked to examine (1) the original and restructured programs' goals, (2) similarities and differences between the new FutureGen and other DOE CCS programs, and (3) if the restructuring decision was based on sufficient information.

GAO reviewed best practices for making programmatic decisions, FutureGen plans and budgets, and documents on the restructuring of FutureGen. GAO contacted DOE, industry partners, and experts.

WHAT GAO RECOMMENDS

GAO recommends that DOE reexamine its restructuring decision, based on the comparative costs, benefits, and risks of the original and restructured programs, as well as other incremental options for modifying the original program. DOE provided technical comments but did not comment on the report's recommendations.

WHAT GAO FOUND

The original FutureGen program and the new restructured FutureGen program attempt to use CCS at coal-fired power plants to achieve near-zero CO_2 emissions and to make CCS economically viable. However, they take different approaches that could affect CCS's commercial advancement. First, the original program aimed at developing knowledge about the integration of IGCC and CCS at one plant; in contrast, the new program could provide opportunities to learn about CCS at different plants, such as conventional ones that use pulverized coal generating technology. Second, the original program was operated by a nonprofit consortium of energy companies at one plant, while the new program called for CCS projects at multiple commercial plants.

The new, restructured FutureGen differs from most DOE CCS programs. The new FutureGen would develop and integrate multiple CCS components at coal-fired plants (including CO_2 capture, transportation, and storage underground). Other programs concentrate on only one CCS component and/or a related component (e.g., capture or capture and compression). However, Round III of DOE's Clean Coal Power Initiative (CCPI) is a cost-shared partnership with industry that funds commercial CCS demonstrations at new and existing coal-fired plants. The new FutureGen is most like CCPI in that both fund CCS commercial demonstrations at several plants to accelerate CCS deployment and require that participants bear 50 percent of the costs, but DOE expects the new FutureGen to have more funding for commercial demonstrations than CCPI. Moreover, the new FutureGen targets a higher amount of CO_2 to be captured and stored (at least 1 million metric tons of CO_2 annually per plant) than CCPI (300,000 metric tons).

Contrary to best practices, DOE did not base its decision to restructure FutureGen on a comprehensive analysis of factors, such as the associated costs, benefits, and risks. DOE made its decision, largely, on the conclusion that costs for the original FutureGen had doubled and would escalate substantially. However, in its decision, DOE compared two cost estimates for the original FutureGen that were not comparable because DOE's $950 million estimate was in constant 2004 dollars and the $1.8 billion estimate of DOE's industry partners was inflated through 2017. As its restructuring decision did not consider a comprehensive analysis of costs, benefits, and risks, DOE has no assurance that the restructured FutureGen is the best option to advance CCS. In contrast to the restructuring decision, DOE's Office of Fossil Energy had identified and analyzed 13 options for incremental, cost-saving changes to the original program, such as reducing the CO_2 capture requirement. While the Office of Fossil Energy did not consider all of these options to be viable, it either recommended or noted several of them for consideration, with potential savings ranging from $30 million to $55 million each.

ABBREVIATIONS

Alliance	FutureGen Industrial Alliance
Btu	British Thermal Units
CCPI	Clean Coal Power Initiative
CCS	carbon capture and storage
CO_2	carbon dioxide
DOE	Department of Energy
EOR	enhanced oil recovery
FE	Office of Fossil Energy
IGCC	Integrated Gasification Combined Cycle
IPCC	Intergovernmental Panel on Climate Change
NEPA	National Environmental Policy Act of 1969
NETL	National Energy Technology Laboratory
syngas	synthesis gas

February 13, 2009
The Honorable Bart Gordon

Chairman
Committee on Science and Technology
House of Representatives

The Honorable Brian Baird
Chairman
Subcommittee on Energy and Environment
Committee on Science and Technology
House of Representatives

The Honorable Daniel Lipinski
House of Representatives

The Honorable Jerry Costello
House of Representatives

Key scientific assessments have underscored the urgency of reducing carbon dioxide (CO_2) emissions to help mitigate climate change. Given the United States' heavy reliance on coal-fired power plants, which emit significant quantities of CO_2, many of these scientific assessments have cited carbon capture and storage (CCS), a developing technology, as a crucial component of any strategy for addressing climate change. CCS involves separating CO_2 from other gases emitted in power plants; capturing the CO_2; compressing it into a liquid form; transporting it (for example, by pipeline) to suitable locations; injecting it into deep underground geologic formations, such as depleted oil reservoirs and saline formations, for long-term storage; and finally, monitoring the presence of the CO_2 at the storage site for a long period of time. Developing CCS is particularly important since total world CO_2 emissions are expected to increase significantly in the near future as the United States continues to use its large coal reserves and as rapidly developing countries, such as China and India, increasingly rely on coal to generate electricity. To date, however, CCS has not been demonstrated on a commercial scale at a power plant, although key stakeholders, such as the International Energy Agency, an organization that advises 28 member countries on energy policy, have noted the importance of commercial- scale demonstration projects for advancing the technology's widespread commercialization.

In 2003, the Department of Energy (DOE) announced its FutureGen program as a $1 billion venture, partnering with the electric power industry to design, build, and operate the world's first coal-fired, zero- emissions power plant. In 2005, the FutureGen Industrial Alliance (Alliance), a nonprofit consortium of some of the largest coal producers and electric power companies in the world, formed to join DOE in this effort. The Alliance agreed to fund 26 percent of the program, and DOE agreed to fund the remaining 74 percent—of which DOE anticipated receiving funding contributions for about 8 percent of the program's total cost from foreign government partners. The agreement was subject to renegotiation and renewal or continuation by both DOE and the Alliance at various stages. In addition to FutureGen, DOE has other clean coal programs with CCS components. For example, Round III of the Clean Coal Power Initiative (CCPI) seeks cost-shared partnerships with industry to fund commercial CCS demonstration at coal-fired power plants.

FutureGen was originally conceived as a research and development project to integrate CCS with another developing technology—integrated gasification combined cycle (IGCC)—in a single power plant at commercial scale.[1] In IGCC power plants, coal is gasified to produce a synthesis gas (syngas), consisting primarily of hydrogen, carbon monoxide, and CO_2. In a process called precombustion CCS, the CO_2 is removed and separated from the syngas before the gas is burned in a combustion turbine to generate electricity. Through IGCC, electricity is generated more efficiently than through conventional pulverized coal-fired technology, the process most widely in use, because IGCC uses less coal to generate the same amount of electricity. In addition, oxygen-fired IGCC plants produce CO_2 as a concentrated gas stream at high pressure that may be captured and stored more easily and cheaply than CO_2 from a typical pulverized coal-fired power plant, which emits CO_2 that must be separated from other gases before storing. Construction on FutureGen was scheduled to begin in 2009, and operations were to begin in 2012. In that year, FutureGen was to begin capturing, storing, and monitoring the stored CO_2 for 3 to 5 years, and then continue monitoring the stored CO_2 for 2 more years. In addition, the FutureGen plant was being designed to serve as a living laboratory host facility for emerging clean coal research programs, including DOE's ongoing coal research program, to help develop advanced technologies that could (1) improve CCS and IGCC, and (2) advance research in other areas, such as hydrogen fuel cells.

By mid-2007, partly because of cost escalations for building power plants around the world, DOE had become increasingly concerned about potential escalating costs for FutureGen. For example, the price of cement, large quantities of which are required for building power plants, had increased by about 30 percent from 2004 to 2006, and certain labor costs for building power plants had increased by over 25 percent, or almost twice the rate of general inflation, from 2001 to 2007. In October 2007, to address these concerns, DOE began renegotiating its share of program costs with the Alliance. In December 2007, after DOE finished conducting the extensive environmental analyses required by the National Environmental Policy Act of 1969 (NEPA)[2] of four potential sites that took over a year, the Alliance announced that it had selected Mattoon, Illinois, for the location of FutureGen. However, DOE had not yet issued its NEPA Record of Decision.[3]Further, DOE had advised the Alliance not to announce a site selection until the Record of Decision had been issued, as contemplated by the cooperative agreement. Subsequently, in January 2008, DOE announced that it would not continue its cooperative agreement with the Alliance and that it was going to take a different approach to FutureGen. DOE stated that this decision was based on concerns over potential cost escalations and the need to more quickly advance commercial technology.

DOE's new approach—the restructured FutureGen—focuses on demonstrating CCS at multiple new or existing commercial coal-fired power plants that may use IGCC or other types of coal plants, such as existing pulverized coal-fired power plants, which comprise 99 percent of all existing coal-fired power plants in the United States. Under the restructured FutureGen, DOE would fund several projects proposed by industry, including entities such as electric power companies, to add CCS to commercial power plants. The plants would begin using CCS by the end of 2015; and, as planned under the original FutureGen, they would be required to capture, store, and monitor the stored CO_2 for 3 to 5 years, and to continue monitoring the stored CO_2 for an additional 2 years. In June 2008, DOE announced that it anticipated providing up to $1.3 billion for the entire restructured FutureGen program, with certain caps in funding for each individual project. The original FutureGen was a DOE research and development project, but the restructured FutureGen is a DOE commercial demonstration project. Under the Energy Policy Act of 2005, a nonfederal source must generally fund not less than 20 percent of a DOE research and development project and not less than 50 percent of a DOE demonstration and commercial application project—that is, industry partners share more of the costs of demonstration projects.[4] However, with both the original and restructured FutureGen, a nonfederal source must pay at least 50 percent of any demonstration component's cost.[5] In October 2008, DOE received a small number of applications for the restructured FutureGen; however, some of these applications are for proposals outside of the restructured FutureGen's scope. DOE is currently assessing proposals received and had stated it expected to announce a selection of projects by December

2008; however, as of the beginning of February 2009, it had made no decision. DOE requested supplemental information from restructured FutureGen applicants which will be reviewed prior to any selection decision.[6]

While IGCC is a promising technology for generating electricity from coal, currently, coal-fired electricity is almost exclusively generated in existing pulverized coal-fired power plants. In these plants, pulverized coal is combusted in air to boil water, which raises steam that, in turn, is routed to turbines to generate electricity. The CO_2 that results from burning coal is exhausted in the flue gas at atmospheric pressure.

In addition to generating electricity in pulverized coal-fired plants or utilizing IGCC technology, oxyfuel combustion is another developmental technology not yet deployed at a commercial scale that could burn pulverized coal to generate electricity. In oxyfuel combustion plants, coal would be burned in pure oxygen diluted with recycled CO_2 or water. Oxyfuel combustion technology could also be one of the technologies considered by DOE in its CCS research efforts, including the restructured FutureGen.

You asked us to examine (1) the goals of the original and restructured FutureGen programs, (2) the similarities and differences between the restructured FutureGen program and other DOE carbon capture and storage programs, and (3) the extent to which DOE used sufficient information to support its decision to restructure the FutureGen program.

In conducting our work, we reviewed FutureGen appropriations, cost estimates, budget justifications, and other DOE documents, including the cooperative agreement and proposed terms for renegotiating the agreement between DOE and the Alliance. We also met with officials from DOE's Office of Fossil Energy (FE), including the National Energy Technology Laboratory (NETL) and the Office of Clean Coal, in addition to officials from the Alliance. We conducted semi-structured interviews with 14 knowledgeable stakeholders from the electric power and coal industries, nonprofit research organizations, academia, and others to determine, among other things, the potential benefits of and key differences between the original and restructured FutureGen programs. We also reviewed public responses to DOE's request for information about the restructured FutureGen and its funding announcement. Finally, we reviewed our recent work and guidance on best practices for cost estimation, program management, and programmatic decision making. A more detailed description of our scope and methodology is presented in appendix I.

We conducted this performance audit from June 2008 to February 2009, in accordance with generally accepted government auditing standards. Those standards require that we plan and perform the audit to obtain sufficient, appropriate evidence to provide a reasonable basis for our findings and conclusions based on our audit objectives. We believe that the evidence obtained provides a reasonable basis for our findings and conclusions based on our audit objectives.

RESULTS IN BRIEF

The overall goals of the original and restructured FutureGen programs are similar in that both programs aim to produce electricity from coal with near-zero emissions by using CCS and to make that process economically viable for the electric power industry. However, the two programs would take different approaches to achieving their goals resulting in, according to knowledgeable stakeholders, two largely distinct programs that could affect aspects of the commercial advancement of CCS differently in the following ways:

- *Type of information gained.* The original FutureGen aimed at developing knowledge about the integration of IGCC and CCS at one power plant. In contrast, the restructured FutureGen could provide opportunities to learn about CCS at different types of coal-fired power plants because the program would be open to coal-fired plants utilizing technologies other than IGCC, such as conventional pulverized coal and oxyfuel combustion. However, under the restructured program, learning about the integration of IGCC and CCS is only possible if DOE receives applications proposing IGCC and selects one for funding.
- *Speed of widespread commercialization of CCS.* It is unclear whether the original FutureGen program or the restructured program would advance the broader roll out of CCS across all of industry more quickly. According to DOE documents, the restructured program is to begin deploying CCS at one or more commercial facilities that generate power for sale in 2015, approximately 5 years earlier than the original program's commercial operations could begin. However, the original program would have begun generating, if not marketing, electricity in 2012. Also, unlike the restructured program, the original program through the Alliance would have included a wide variety of industry and international partnerships, thereby fostering widespread commercialization of CCS technology and the use of that technology.

- *Testing advanced technology*. The original FutureGen would have served as a living laboratory host facility for emerging technologies, aimed at the goal of near-zero emissions (such as hydrogen fuel cells), and for gaining broad industry acceptance for these technologies. In contrast, the restructured FutureGen would not include a facility for testing these technologies, and its ability to advance them would, therefore, be limited.

The restructured FutureGen differs in important ways from most of DOE's other CCS programs, with the exception of one program—Round III of CCPI. Both the restructured FutureGen and CCPI (1) fund the commercial demonstration of CCS at coal-fired power plants, and (2) require industry participants to bear at least 50 percent of costs. The restructured FutureGen would potentially have more available funding for commercial demonstrations than CCPI, and the restructured FutureGen targets a higher amount of CO_2 to be stored (at least 1 million metric tons of CO_2 stored annually, per plant) than CCPI (300,000 metric tons of CO_2 stored or put to use annually, per plant). However, because CCPI's goals may be more realistic or attainable for commercial partners than those of the restructured FutureGen, CCPI may receive more proposals and, hence, more industry participation. Regarding the restructured program's differences from most of the other CCS programs, the restructured FutureGen would integrate key components of CCS at commercial coal- fired power plants, such as CO_2 capture, compression, transport, storage, and monitoring of stored CO_2; in contrast, most of DOE's other CCS programs concentrate on developing individual components of CCS, such as CO_2 storage, and/or an individual component and a related one, such as capture and compression.

DOE did not use sufficient information to support its decision to restructure FutureGen. According to our recent work and best practices, a decision to terminate or significantly restructure an ongoing program should be informed by timely and sufficient information on the costs, benefits, and risks of such a decision.[7] DOE did not prepare a comprehensive analysis of the costs, benefits, and risks of its decision to replace the original FutureGen with the restructured program. DOE made its decision based, in large part, on its conclusion that construction and material costs for the original program would continue escalating substantially in the indefinite future and that life-cycle costs were likely to double. However, according to economic forecasting organizations, such as DOE's Energy Information Administration, significant cost escalations for building power plants, in general, do not typically continue in the long run. Also, DOE reached this conclusion by comparing its cost estimate for the original FutureGen ($950 million in constant 2004 dollars) with the Alliance's 2006 estimated life-cycle costs for the program through 2017 (about $1.8 billion, considering inflation). In explaining his decision to restructure FutureGen, the Secretary of Energy noted that the projected program cost had "nearly doubled," from $950 million to $1.8 billion. However, that assertion did not take into account a major difference between the two estimates: one was based on constant dollars and the other on inflated dollars. Our analysis indicates that the Alliance's estimate in constant 2005 dollars would be approximately $1.3 billion—an increase of about $370 million, or about 39 percent, over DOE's estimate, not a near doubling of costs. As DOE's restructuring decision was not based on a comprehensive analysis of the associated costs, benefits, and risks, DOE has no assurance that the restructured program is the best option to accomplish the goal of promoting the accelerated and widespread commercial advancement of CCS. In contrast to the restructuring decision, FE identified and analyzed 13 other options for incremental, cost-saving changes to the original program, such as reducing the CO_2 capture requirement. While FE did not consider all of these options to be viable, it either recommended or noted several of them for consideration, each with potential savings from $30 million to $55 million.

To help ensure the widespread commercial advancement of CCS while protecting taxpayer interests, we are recommending that, before implementing significant changes to FutureGen or before obligating additional funds for such purposes, the Secretary of Energy should direct DOE staff to prepare a comprehensive analysis comparing the relative costs, benefits, and risks of a range of options, including the original and restructured FutureGen programs and incremental options for modifying the original program.

In commenting on a draft of this report, DOE thanked us for the opportunity to review the draft. In its response, DOE did not provide comments on the report's conclusions or recommendations. However, DOE provided us with technical comments, which we have incorporated into the report, as appropriate. See appendix III for DOE's comments and our response to these comments.

BACKGROUND

Global emissions of greenhouse gases, such as CO_2, from human activities have grown markedly since preindustrial times—since about the year 1750—with an increase of 70 percent from 1970 to 2004. Most scientists agree that increased greenhouse gases in the atmosphere are the primary cause of the rise in global temperatures in recent decades. The Intergovernmental Panel on Climate Change (IPCC) expects greenhouse gas emissions to continue to increase over the next few decades, resulting in a continued rise in global temperatures and related harmful impacts, including the flooding of large populated coastal areas, a reduction in the production of some crops and livestock productivity, and a decrease in the availability of fresh water in certain parts of the world.

According to the National Academy of Sciences, CO_2 levels in the atmosphere are currently at their highest in at least 650,000 years and are continuing to rise. Global increases in concentrations of CO_2 in the atmosphere are due primarily to the burning of fossil fuels—such as petroleum and coal—for energy, industrial processes, and transportation. Coal is currently the world's leading source of electricity, and the use of coal to generate electricity around the world is projected to double by 2030. Coal-fired power plants provide about one-half of the supply of electricity used in the United States, and DOE's Energy Information Administration estimates that coal accounts for approximately one-third of the total CO_2 emissions in the United States. In addition, the International Energy Agency anticipates that the two largest developing countries— China and India—will drive increased demand for coal to meet their growing electricity needs. According to the International Energy Agency, these countries' heavy reliance on coal has already contributed significantly to recent increases in global CO_2 emissions.

To prevent the anticipated increase in coal-fired electricity generation from emitting significant amounts of CO_2 into the atmosphere, many are suggesting CCS as a tool that allows for continued coal use, while mitigating its effect on the climate.[8] CCS comprises several components: separating CO_2 from other gases emitted by the plant; capturing emitted CO_2; compressing the CO_2 into a fluid state to facilitate its transportation; transporting it to a storage location; injecting the CO_2 into geologic formations, such as depleted oil and gas reservoirs and saline formations, for storage; and monitoring the storage site to verify that the CO_2 remains in place. A successful CCS system must integrate all of these components.

Currently, three major approaches have been identified for capturing CO_2 at coal-fired power plants: (1) generating electricity using pulverized coal as a fuel in conventional power plants with postcombustion capture of CO_2, (2) generating electricity using IGCC technology with precombustion capture of CO_2, and (3) using pulverized coal as a fuel in power plants that utilize oxyfuel combustion technology to generate power and capture CO_2.

- *Postcombustion capture and pulverized coal-fired power plants*: Pulverized coal-fired power plants, which comprise 99 percent of all existing coal-fired power plants in the United States, burn pulverized coal to boil water, which raises steam that, in turn, is routed to turbines to generate electricity. The CO_2 that results from burning coal is exhausted in the flue gas at atmospheric pressure and a concentration of 10 to 15 volume percent. Postcombustion capture of CO_2 occurs after the coal is burned. The technology for capturing the CO_2 could be retrofitted onto existing power plants. However, according to DOE, the postcombustion capture of CO_2 is a challenging approach because (1) the low pressure and dilute concentration dictate a high actual volume of gas to be treated; (2) trace impurities in the flue gas tend to reduce the effectiveness of the CO_2 adsorbing processes; and (3) compressing captured CO_2 from atmospheric pressure to pipeline pressure (about 1,200 to 2,000 pounds per square inch) requires a large amount (an estimated 20 to 40 percent) of the electric power generated by the power plant.[9]
- *Precombustion capture and IGCC power plants*: This approach would be used at coal plants that use IGCC, a technology for generating electricity that has been deployed on a commercial scale at only two coal-fired power plants in the United States. In an IGCC plant, coal is gasified through a thermochemical process to break it down into its chemical constituents and produce a synthesis gas (syngas) which consists mostly of hydrogen, carbon monoxide, and CO_2.[10] The syngas is then treated to remove contaminants, such as ammonia and mercury, and burned in a combustion turbine to generate electricity. Precombustion capture in IGCC plants can occur because the CO_2 in the syngas is at a very high pressure, which allows it to be captured before the combustion of the syngas.
- *Oxyfuel combustion*: Oxyfuel combustion, which is in its developmental stages, is a technology that is being developed for using coal to generate electricity that could reduce CO_2 emissions.

> According to DOE, oxyfuel combustion could be applied to existing pulverized coal-fired plants. Oxyfuel combustion burns coal using pure oxygen diluted with recycled CO_2 or water. As a result, oxyfuel combustion emits primarily CO_2 and water vapor, with some excess oxygen, facilitating the capture of the CO_2 by condensing the water in the exhaust stream. Because separating out the CO_2 is not necessary under this approach, the CO_2 capture consists essentially of drying and compressing the CO_2. However, depending on the level of excess oxygen and other trace components, some additional gas cleanup may be required to make the CO_2 suitable for transportation.

After being captured, the CO_2 would be transported, likely via pipeline, to a storage site and injected at depths of over 800 meters (or about 2,600 feet) into underground geologic formations (such as depleted oil reservoirs and saline formations), thought to be conducive for isolating the CO_2 for hundreds to thousands of years. We reported in 2008 that among the barriers to CCS deployment are regulatory and legal uncertainties regarding the liability for CO_2 leakage and the ownership of CO_2, once injected. Once injected, the CO_2 must be monitored to ensure it does not escape into the environment.[11]

On February 27, 2003, the President announced FutureGen as a cost-shared project between DOE and industry to create the world's first coal-fired, zero emissions electricity and hydrogen production power plant. The production of hydrogen was to support the President's Hydrogen Fuel Initiative to create a hydrogen economy for transportation. The original FutureGen plant was planned to operate at a commercial scale as a 275 megawatt IGCC facility that would capture and store at least 1 million metric tons of CO_2 per year. In December 2005, DOE signed a cooperative agreement with the nonprofit Alliance.[12] Pursuant to the agreement, the Alliance was to design, construct, and operate the FutureGen plant, and DOE was to provide project oversight, conduct the environmental analyses required by NEPA, and coordinate the participation of foreign governments.[13] The project was to run through November 2017 and operate as the cleanest fossil fuel-fired power plant in the world. After completion of the formal project, the FutureGen plant was expected to continue operating for the typical lifespan of a power plant—usually 30 to 50 years—generating electricity and providing a platform for energy research. On January 30, 2008, DOE announced that it had decided to take the FutureGen program in a different direction. DOE stated that it would demonstrate CCS at multiple commercial-scale power plants, including retaining the integration of CCS and IGCC. DOE referred to this new approach as the restructured FutureGen program. In June 2008, in a funding announcement for the restructured program, DOE stated that it expected it would have about $290 million available through fiscal year 2009 for its share of funding for the program. (See app. II for an overview of DOE budget authority and obligations for FutureGen.)

The Goals of the Original and Restructured FutureGen Programs Are Largely Similar, but the Programs' Different Approaches May Lead to Different Results

The overall goals of the original and restructured FutureGen programs are largely similar in that both programs seek to produce electricity from coal with near-zero emissions by using CCS, and to make that process economically viable for the electric power industry. However, the programs outline different approaches for achieving their goals, which could affect the commercial advancement of CCS differently in several ways.

With a Few Key Exceptions, the Goals of the Original and Restructured Programs Are Largely Similar

Both the original and restructured programs aim to establish the feasibility and economic viability of producing electricity from coal with near-zero emissions by employing CCS. The programs' goals for storing CO_2 and limiting other emissions, such as mercury and sulfur, are also similar; except that the requirement for the amount of carbon to be captured has been reduced from 90 percent in the original program to 81 percent in the restructured program (see table 1).

Knowledgeable stakeholders told us that this decrease in carbon capture is of modest significance and that a goal of 81 percent is still very ambitious and costly. DOE received similar feedback in responses to its request for information from the public about its plan to restructure FutureGen. Eighteen of the 49 respondents indicated that the 90 percent goal would be too restrictive for industry participants because of the additional energy required to capture and compress CO_2, often referred to as the energy penalty, and the fact that no power plant to date has been designed to operate with the conditions necessary to achieve a 90 percent capture. Some respondents suggested setting a lower capture goal, such as 65 percent, and adopting an incremental approach over time to reach 90 percent that would eventually allow industry to obtain baseline data and demonstrate reliability and widespread confidence in CCS. One respondent wrote that the "90% capture level was appropriate for a 'living laboratory' like the originally proposed FutureGen project but a 30% level is most suitable for a commercial IGCC facility."

Both the original and restructured programs would operate plants at a commercial size.[14] However, the restructured program, a DOE commercial demonstration project, seeks to accelerate the commercial deployment of CCS (that is, generating and selling electricity to earn profits) by implementing CCS at one or more commercial facilities by 2015— approximately five years earlier than the original program's commercial operations could begin. The original program, a DOE research and development project, would begin generating electricity in 2012, a few years earlier than the restructured FutureGen; but, it could not begin operating as a profit-seeking commercial facility until after the nonprofit Alliance sells it, which is currently anticipated to occur in 2020. Knowledgeable stakeholders told us that the restructured program's time line for the commercial deployment of its project(s) might be ambitious because legal and environmental issues related to siting and permitting, in particular for CCS, could slow implementation. They also stated that the required NEPA analyses, which must be completed prior to beginning construction, could take up to 3 years. In contrast, DOE had completed its NEPA analyses for the original FutureGen. Moreover, the governments of the two states—Texas and Illinois—where the four finalist sites for the original FutureGen were located, had agreed to assume liability for the injected CO_2. DOE officials told us that, unlike the original program, a primary goal of the restructured FutureGen was to facilitate the siting and permitting process for CCS by implementing multiple projects in different locations.

The Different Approaches for Achieving Goals Could Have Different Impacts on the Commercial Advancement of CCS

Because of the different approaches for achieving their goals, the original and restructured FutureGen programs could have different impacts on the commercial advancement of CCS (see table 2).

Table 1. Storage and Emissions Goals for the Original and Restructured FutureGen Programs

Storage and emissions goals	Original FutureGen	Restructured FutureGen
Carbon	Capture at least 90%[a]	Capture at least 81%[a]
CO_2	Store at least 1 MMT/year[b]	Store at least 1 MMT/year[b]
Sulfur	Remove at least 99%	Remove at least 99%[c]
Mercury	Remove at least 90%	Remove at least 90%[d]
Btu NOx	Reduce to less than .05 lb/million	Reduce to less than .05 lb/million
Btu particulate matter	Reduce to less than .005 lb/million	Reduce to less than .005 lb/million

Source: GAO analysis of DOE program documents.

[a] The original FutureGen program's goal is to capture 90 percent of the CO_2 in the total plant gas stream, while the restructured FutureGen program requires capture of 81 percent of the carbon in the total plant gas stream. According to DOE officials, the carbon capture goal of the restructured program includes the capture of CO_2, as well as other carbon-based gases—such as methane (CH_4), another greenhouse gas. DOE officials explained that they changed the measure from CO_2 to carbon for clarification purposes and because the restructured FutureGen was designed to accommodate a range of gasification and combustion technology configurations. Also, while the goal for the restructured FutureGen states 90 percent carbon capture, the minimal performance requirement is 81 percent of carbon.

[b]Per demonstration unit or plant.

[c]Based on the sulfur content of the coal, or less than 0.04 lbs/million Btu if there is low sulfur concentration in the coal.

[d]Based on the mercury content of the coal.

The type of information gained from the programs may vary. First, the original program would have developed knowledge about CCS at IGCC plants, while the restructured program could allow for opportunities to learn about CCS at both IGCC and other types of coal plants. Knowledgeable stakeholders whom we contacted stated that DOE could benefit by taking advantage of the opportunity under the restructured FutureGen program to learn about CCS at multiple types of plants. They explained that opportunities to learn from multiple plant sites in different regions with various technologies would provide a wide range of knowledge about the implementation of CCS in various contexts.

Table 2. Key Differences in the Approaches of the Original and Restructured FutureGen Programs

Original FutureGen	Restructured FutureGen as currently designed
Only includes IGCC.	May or may not include IGCC; IGCC was identified as a goal, but its inclusion in the program depends on the applications received and selected.
Accelerated deployment of CCS at commercial facilities is not a goal.	Accelerated deployment of CCS at commercial facilities is a goal.
Includes a living laboratory host facility for advanced technologies, such as fuel cells.	Does not include a living laboratory host facility for advanced technologies.
Project would be operated by a nonprofit consortium of 11 industry partners (FutureGen Industrial Alliance).	Each project could be operated by a nonprofit or for-profit entity.
Research and development project (DOE cost share capped at 74%).	Demonstration project (DOE cost share capped at 50%).
Consists of one plant.	Potential for multiple sites, depending on the applications received and selected.
Includes international involvement.	No international involvement.

Source: GAO analysis of DOE program documents.

Similarly, 30 of 49 respondents to DOE's request for information about the restructured program indicated that it would be beneficial if the restructured program were to include both IGCC and other types of coal plants. In addition to other organizations, such as the National Academy of Sciences, we have noted that the benefits of learning about CCS technologies are also applicable to existing pulverized coal-fired plants, since they account for an overwhelming share (about 99 percent) of the world's coal-fired power plants.[15] However, one of the intended benefits of the restructured program—providing opportunities to learn from multiple plants about various technologies—may not be fully realized since DOE received only a small number of applications. If an application for IGCC has not been received or is not selected, the loss of an IGCC plant with integrated CCS capability is significant because, according to the draft strategic planning document for the restructured program, demonstrating this technology is a key solution for reducing atmospheric CO_2 emissions from coal-fired power plants. Comments submitted to DOE and knowledgeable stakeholders we interviewed indicated that the carbon capture goal for the restructured program was too restrictive for commercial facilities. One stakeholder stated that the restructured program goals might be overly optimistic about what commercial projects are willing to do. As a result of receiving only a small number of applications, the restructured program is not as likely to develop as broad a base of knowledge as it could have if more applications were received.

Second, it is unclear whether the original FutureGen program or the restructured program could have advanced the broader roll out of CCS more quickly across all of industry. According to DOE documents, the restructured program is intended to begin deploying CCS at one or more commercial facilities in 2015, approximately five years earlier than the original program's commercial operations (that is, generating and selling electricity) could begin. The original program, a DOE research and development project, would have begun generating electricity in 2012, a few years earlier than the restructured FutureGen, but it could not have begun operating as a profit-seeking commercial facility until after the nonprofit Alliance sold it, which was anticipated to occur in 2020. Moreover, unlike the restructured program, the original FutureGen would have included a wide variety of industry partners (including foreign government partners, which are absent from the restructured program). In addition, more industry partners could have joined the Alliance and its 11 members over the course of the original program. As a result of its wider participation, the original FutureGen could potentially have advanced the broader roll out of CCS across all of industry and internationally, instead of at only a few commercial facilities, more quickly than the restructured program. DOE officials told us that the original program would likely improve the global advancement of CCS more quickly than the restructured program due to its various international partnerships. They stated that DOE is developing an approach to recoup the loss of international involvement that resulted from restructuring FutureGen.

Third, the restructured program will not serve as a living laboratory host facility for technologies emerging from energy research and development programs aimed at the goal of near-zero emissions and for gaining broad industry acceptance for these technologies. The original FutureGen plant was to be designed with the ability to test various technologies that are scalable to full size, such as fuel cells, advanced gasification, and membrane air separation systems. Without the opportunity to test these emerging research and development technologies, the restructured FutureGen might result in a slower advancement of CCS than the original program may have yielded. According to the cooperative agreement between DOE and the Alliance, emerging technologies, such as fuel cells, could have been tested at the original program's living laboratory host facility. In a September 2007 presentation to DOE's Deputy Secretary, NETL noted the impact of removing the living laboratory, saying it would "significantly delay the availability of the technology for commercial deployment" and have a "significant programmatic impact." DOE officials told us that they have not yet determined where these technologies will be tested.

THE RESTRUCTURED FUTUREGEN DIFFERS FROM MOST OF THE OTHER DOE CARBON CAPTURE AND STORAGE PROGRAMS, BUT IT IS SIMILAR TO CCPI IN SEVERAL WAYS

DOE manages a portfolio of clean coal programs that research and develop CCS technology or demonstrate its application. Focusing on commercial coal-fired power plants, the restructured FutureGen would integrate key components of CCS, such as CO_2 capture, compression, transport, storage, and monitoring of CO_2 at the storage location. However, the restructured FutureGen is similar in some ways to Round III of CCPI, but CCPI's goals are more modest than those of the restructured FutureGen and, hence, may be more achievable for industry partners. The other CCS programs include the (1) Regional Carbon Sequestration Partnerships, (2) Innovations for Existing Plants Program, (3) Advanced Turbines Program, (4) Advanced Integrated Gasification Combined Cycle Program, and (5) Round III of the Title 17 Incentives for Innovative Technologies Loan Guarantee Program (Loan Guarantee Program). Four of these five CCS programs do not integrate all key components of CCS and concentrate on developing one or two related components of CCS, such as CO_2 separation, CO_2 storage, or CO_2 capture with related compression.

The Restructured FutureGen Program Is Similar to Round III of CCPI in Several Ways, but CCPI's Goals May be More Achievable for Industry Partners

Both the restructured FutureGen and CCPI are cost-shared partnerships with industry, in which DOE funds no more than 50 percent of the costs. Like the restructured FutureGen, Round III of the CCPI program funds the commercial demonstration of CCS at coal-fired power plants.[16] Round III of

CCPI seeks to demonstrate, at a commercial scale, advanced coal-based technologies that capture and store carbon, or put CO_2 emissions to beneficial reuse, such as to enhance oil recovery.[17] The proposals for Round III of CCPI were due to DOE by January 15, 2009, and DOE expects to announce its selections in July 2009.

In public comments on DOE's request for information and the draft funding announcement for the restructured FutureGen, two respondents noted the similarity between the restructured FutureGen and CCPI. They suggested that DOE explain the linkages and possibly combine the programs. However, important differences exist in the two programs' goals. First, while both programs have annual requirements for the capture of CO_2 emissions, the restructured FutureGen requires 1 million metric tons of CO_2 per plant, while CCPI requires 300,000 metric tons of CO_2 per plant. Knowledgeable stakeholders told us that CCPI's goal of capturing 300,000 metric tons of CO_2 is more realistic and attainable by commercial facilities than the restructured FutureGen's goal of 1 million metric tons. Another noteworthy distinction is that the restructured FutureGen requires the 1 million metric tons of CO_2 emissions to be stored in saline formations, whereas the 300,000 metric tons of CO_2 emissions that CCPI requires to be captured can either be stored or be put to beneficial reuse, such as to enhance oil recovery. The latter opens up more options for industry partners and can serve as an attractive opportunity for increasing revenue in the project by selling the CO_2. Finally, because CCPI's goals may be more realistic or attainable for commercial partners, more proposals may be submitted to CCPI than the restructured FutureGen. For example, two officials from electric utility companies said that, despite the potentially greater amount of funding available through the restructured FutureGen ($1.3 billion, subject to future appropriations) than CCPI ($440 million, subject to future appropriations), their companies would apply for CCPI over the restructured program because they could meet CCPI's goals.

The Restructured FutureGen Differs from Most Other DOE CCS Programs

The restructured FutureGen and other DOE CCS programs strive to reduce CO_2 emissions by advancing CCS. However, while most of these programs do not integrate all key components of CCS, the restructured FutureGen integrates all key components of CCS. The other CCS programs include the (1) Regional Carbon Sequestration Partnerships, (2) Innovations for Existing Plants Program, (3) Advanced Turbines Program, (4) Advanced Integrated Gasification Combined Cycle Program, and (5) Round III of the Title 17 Incentives for Innovative Technologies Loan Guarantee Program.

The Regional Carbon Sequestration Partnerships seek to develop the technology, infrastructure, and regulations necessary to implement CO_2 storage.[18] The 7 regional partnerships are composed of over 350 organizations, 42 states, 4 Canadian provinces, and 3 Native American tribes. Now entering Phase III,[19] the regional partnerships are working to implement 7 large-scale projects that will demonstrate the long-term, effective, and safe storage of CO_2 in the major underground geologic formations throughout the United States and portions of Canada. The CO_2 stored through the projects can come from coal-fired power plants or other sources, such as ethanol production plants. The injection of CO_2 into geologic formations will continue over several years, and the monitoring will continue through 2017.

The Innovations for Existing Plants Program focuses on developing CO_2 capture and compression technologies to assist existing coal-fired power plants.[20]Through this program, DOE is providing $36 million in funding for 15 projects to develop new and cost-effective CO_2 capture technologies for existing power plants. According to DOE, all 15 projects selected have received funding. The projects will be implemented across 11 states and will last for 2 to 3 years. Projects will focus on five areas of interest for CO_2 capture: membranes, solvents, sorbents, oxyfuel combustion, and chemical looping.

The Advanced Turbines Program focuses on creating the technology base for turbines that will permit the design of IGCC plants with CCS that can operate at near-zero emissions, thereby facilitating CO_2 capture. According to DOE, the development of new turbines technology could improve applications of IGCC by reducing the costs of producing electricity from coal.

Similarly, the Advanced Integrated Gasification Combined Cycle Program also focuses on one aspect of CCS—developing gasification technology to enable CO_2 capture. The program aims to develop advanced gasification technologies to enable CO_2 capture with minimal impact on the cost of electricity. DOE reports that by 2012, gasification technology will be integrated at pilot scale with CO_2 separation, capture, and sequestration into near-zero atmospheric emissions configurations that can, ultimately, provide electricity with less than a 10 percent increase in cost.

Table 3. DOE Programs Supporting Carbon Capture and Storage

	Restructured FutureGen	Clean Coal Power Initiative (Round III)	Regional Carbon Sequestration Partnerships
Description	Demonstration of capture and storage of CO_2.	Demonstration of capture and storage, or beneficial reuse, of CO_2.	Demonstration of CO_2 storage in geologic formations.
Integrates all key CCS components[a]	Yes	Yes	No[b]
Commercial site	Yes	Yes	No
Demonstration or R&D	Demonstration	Demonstration	R&D
DOE cost share	50%[c]	50%[c]	80%[d]
Carbon storage required (amount and location)	Yes, storage required. At least 1 million metric tons/year of CO_2 must be	No, storage not required. 300,000 metric	Yes, storage required. Some will store up
	stored in a saline forma-tion and any excess of 1 million metric tons/year can be put to beneficial reuse, such as for enhanced oil recovery.	tons/year can either be stored or put to beneficial reuse, such as for enhanced oil recovery.	to 1 million metric tons/year in geol-ogic formations.
Develops CO_2 capture and compression technologies for pulverized coal power plants, which represent the majority of existing coal plants.	Creation of new turbines for IGCC plants that will include CCS and facilitate near-zero atmospheric emissions.	Supports the development of advanced gasification technologies to enable CO_2 capture with minimal impact on the cost of electricity.	Loan guarantees for activities at retrofit-tted and new facilities that incorporate carbon capture and sequestration, other beneficial uses of carbon, or advanced coal gasification.
No	No	No	Possible
No	No	No	Yes
R&D	R&D	R&D	n/a
80%	80%	80%	n/a
n/a	n/a	n/a	n/a

Source: GAO analysis of DOE program documents.

[a]Key components include capture, compression, transport, storage and measurement, monitoring and verification.

[b]The Regional Partnerships will conduct large-scale geological sequestration testing that will require the participants to secure sufficient quantities of CO_2 needed to demonstrate CO_2 storage, monitoring, and verification. However, while DOE will cost share in the acquisition

of CO_2, it will not fund the development and/or testing of CO_2 capture technologies under the Regional Partnership program.

[c]DOE cost sharing is generally capped at 50 percent. According to DOE officials, it is quite common under DOE's commercial demonstration programs for the Government cost share to be well below 50 percent of the total project cost.

[d]DOE cost sharing is generally capped at 80 percent. Private sector cost sharing under the seven Regional Partnerships averages 34 percent.

Finally, Round III of the Title 17 Incentives for Innovative Technologies Loan Guarantee Program will provide up to $8 billion in loan guarantees for energy projects that satisfy three criteria: avoid, reduce, or sequester air pollutants or greenhouse gases; employ new or significantly improved technologies, compared with commercial technologies in service at the time the guarantee is issued; and provide a reasonable prospect of repayment.[21] Initial applications for Round III of the program were due to DOE in December 2008. We recently reported on DOE's progress in (1) issuing final regulations to govern this program, (2) taking actions to help ensure that the program is managed effectively and to maintain accountability, and (3) determining whether there were inherent risks due to the nature and characteristics of this program that may affect DOE's ability to make the program pay for itself and support a broad spectrum of innovative energy technologies.[22] Table 3 summarizes the comparison of DOE programs supporting CCS.

DOE Did Not Support Its Decision to Restructure FutureGen with Sufficient Information on Costs, Benefits, or Risks

According to our recent work and best practices, a decision to terminate or significantly restructure an ongoing program should typically be informed by timely and sufficient information on the costs, benefits, and risks of such a decision. While DOE had reason to be concerned about the escalating costs of the original FutureGen, it made its decision to cancel that program and replace it with the restructured FutureGen based, in large part, on a comparison of cost estimates that were not comparable. That is, it compared one estimate that was in current dollars with one that was in constant dollars. In restructuring FutureGen, DOE did not sufficiently analyze the costs, benefits, and risks of canceling the original FutureGen and replacing it with a significantly restructured program. A comprehensive analysis could have helped DOE determine how the costs, benefits, and risks of the restructured FutureGen compared with those of the original FutureGen. Because it did not conduct such an analysis, DOE cannot be assured that the restructured program is the best option to accelerate the widespread commercial advancement of CCS more quickly than the original program. Other options, rather than dramatically restructuring the program, were possible that could have preserved some of the benefits of the original program, including ensuring the integration of IGCC and CCS at the FutureGen facility. For example, FE identified and analyzed 13 other options for incremental, cost-saving changes to the original program, such as reducing the CO_2 capture requirement. While FE did not consider all of these options to be viable, it recommended or noted several of them for consideration with potential savings from $30 million to $55 million each.

DOE Decided to Restructure FutureGen Based, in Large Part, on a Comparison of Cost Estimates that Were Not Comparable

In January 2007, as part of its initial conceptual design report for the original FutureGen, the Alliance estimated the cost of the original program at about $1.8 billion. The Alliance's report explained that this estimate included inflation through 2017—the last year of the anticipated life of the program—and was the equivalent of almost $1.4 billion in constant 2006 dollars. This report also stated that, after subtracting anticipated revenue from program activities, such as the sale of electricity, the estimate was similar to DOE's 2004 estimate of $950 million. However, DOE officials told us that DOE's estimate did not subtract anticipated revenue. In March 2007, after approving the Alliance's cost estimate, DOE renewed the cooperative agreement with the Alliance to proceed with developing a preliminary design for FutureGen by June 2008, including a revised cost estimate and a risk analysis. The preliminary design was to be based on a specific site and technology for the program—information that has an important impact on the program's overall cost because labor expenses vary from location to location, and technology costs and designs, such as for turbines, vary depending on the specific manufacturer and vendor.

The March 2007 renewed cooperative agreement listed approximately $1.8 billion as the current estimated cost of the project. However, senior DOE officials soon began to express concerns about escalating program costs, and they directed FE officials to develop recommendations for controlling costs. In September 2007, FE officials presented several recommendations for incremental changes to control costs to the Deputy Secretary of Energy; they also noted various measures already in place for controlling costs, such as monthly progress reports and a risk management program. Importantly, none of the recommendations indicated that DOE should cancel the original program and restructure FutureGen; moreover, FE officials told us that they did not prepare any analysis or recommendations for senior DOE officials that resembled what was to become the restructured program.

According to DOE, following this presentation, senior DOE officials directed FE to negotiate with the Alliance new cost sharing arrangements under the cooperative agreement, which was scheduled for continuation in June 2008. The Alliance agreed to meet to renegotiate the terms of the cooperative agreement. Over the course of several meetings, the parties discussed various funding scenarios and exchanged proposed term sheets. Subsequently, however, the Alliance and DOE did not reach agreement. In December 2007, the Alliance sent a letter to DOE stating that it preferred to proceed under the existing cooperative agreement until FutureGen's costs and risks could be assessed with input from the preliminary design report and cost estimate that were due by June 2008.

Also in December 2007, the Secretary of Energy briefed senior presidential advisers that the estimated cost of FutureGen had nearly doubled—from $950 million to $1.8 billion—and that costs were expected to continue rising. In addition, according to the briefing documents, DOE planned to end its partnership with the Alliance and was developing a new strategy for FutureGen that would cap the government's financial exposure. The briefing documents explained that DOE's new approach for FutureGen would fund only the CCS-related technology associated with multiple commercial IGCC plants, rather than the entire construction of a single plant with CCS. Around this time, according to DOE officials, senior DOE officials directed FE to develop the restructured FutureGen program. In response, these officials told us, many high-level offices within DOE collaborated on developing a draft strategic planning document for the restructured program. According to these officials, the draft strategic planning document that they finalized in January 2008 was the first complete document about the restructured FutureGen. On January 30, 2008, DOE publicly announced that it was restructuring FutureGen to provide a ceiling on federal contributions and that the restructured program was a more cost-effective approach. On this same day, DOE notified the Alliance that it was restructuring FutureGen and would not continue its cooperative agreement with the Alliance. DOE informed the Alliance that it was restructuring FutureGen in response to serious concerns over substantial escalation in projected costs, including what the agency concluded would be the likely continued escalation of the costs. DOE officials also stated that they disapproved of the Alliance's decision to announce the selection of a project site before DOE issued its NEPA Record of Decision. According to DOE, prior to the site selection announcement and without knowledge of the Alliance's choice of site, DOE had asked the Alliance not to go forward with the announcement and further advised the Alliance against making an announcement until the Record of Decision had been issued. DOE officials also said that, in their negotiations on measures that could limit DOE's financial exposure, they lost confidence in the ability of the Alliance to fund its share of the project cost.

Although comparing cost estimates can provide valuable insight about the impact of escalating costs on a project, DOE based its decision to restructure FutureGen, in large part, on a comparison of cost estimates that were not actually comparable. That is, in 2004, DOE had estimated that the cost of the original FutureGen would be $950 million in constant 2004 dollars. In contrast, the Alliance's 2007 estimate of about $1.8 billion was in current dollars, which reflected inflation over the course of the program from 2005 through 2017.[23] In explaining his decision to restructure FutureGen to senior presidential advisers, the Secretary of Energy indicated that the projected program costs had "nearly doubled," from $950 million to $1.8 billion. However, comparing constant dollars, which exclude inflation, with current dollars, which reflect inflation, is misleading. Our calculations show that the Alliance's current dollar estimate of roughly $1.8 billion is equivalent to approximately $1.3 billion in constant 2005 dollars—an increase in total program costs of about $370 million, or about 39 percent—not a near doubling of costs.[24]

In addition, the cost estimates by DOE and the Alliance were prepared early in the project and, as a result, were based on conceptual designs for FutureGen, including power plant case studies and a blanket 10 percent increase incorporated into the Alliance's estimate to allow for the first-ofa-kind nature of some of the plant's components and integration issues. However, neither estimate considered costs for specific types of technology or a specific location. If DOE had waited approximately 6 months for the Alliance's technology-specific and site-specific cost estimate, due by June 2008 as part of its preliminary design report, before deciding whether to restructure the program, it would have had the benefit of more current and complete information, including the latest information on escalating costs, when making decisions about how to move forward with FutureGen.[25] In addition, regarding FutureGen's total cost, the March 2007 cooperative agreement stated that DOE and the Alliance recognized that many uncertainties—such as plant design, site selection, and market conditions—still existed in developing a firm cost estimate.[26]

In May 2008, the Secretary of Energy testified before Congress that FutureGen was conceived as a $950 million venture and that its estimated cost had increased to roughly $1.8 billion; however, the Secretary's prepared statement did not indicate that the first estimate was in constant dollars, while the second was in current dollars.[27] The Secretary also testified that DOE believed its costs would continue to escalate. We requested that DOE provide us with the analysis that supported DOE's anticipated escalation. In October 2008, DOE officials told us that the ongoing cost escalations were unprecedented and that they had looked internally across various indexes, including the Bureau of Labor Statistics, to get a sense of prospective escalation. However, they stated that they did not have any written or comprehensive analysis. They added that they did not prepare a position paper, study, or generate any analysis examining current or future escalation for the decision to restructure FutureGen. Moreover, economic forecasting organizations, such as DOE's Energy Information Administration, have found that significant cost escalations, such as those for building power plants over the past several years, do not typically continue in the long run.[28]

A Comprehensive Analysis Could Have Helped DOE Determine How the Costs, Benefits, and Risks of the Restructured FutureGen Compared with Those of Other Options

DOE did not prepare a comprehensive analysis comparing the relative costs, benefits, and risks of the original and restructured FutureGen programs before making the decision to replace the original program with the restructured FutureGen. On two different occasions, DOE officials told us that the agency did not prepare such an analysis. These officials told us that the Secretary of Energy's May 2008 congressional testimony included the agency's official explanation for why it decided to restructure FutureGen. In September 2008, we asked DOE to provide us with additional information, including the agency's official position on why it decided to restructure FutureGen, all the factors upon which DOE based the decision, the extent to which the decision was based on documented supporting analysis, and a copy of any such analysis. In January 2009, after we sent a draft of this report to DOE for review and comment, DOE responded to our request for additional information, stating that the detailed analysis supporting its decision to restructure FutureGen could be found in the draft strategic planning document for the restructured program and that this document discussed the factors considered by DOE in making the decision to restructure FutureGen. However, as previously discussed in our findings, the draft strategic planning document was not completed in time to inform the decision to restructure FutureGen. In addition, we do not consider the draft strategic planning document to be comprehensive because it did not assess:

- whether costs for the original FutureGen would escalate substantially in the future;
- the relative costs, benefits, and risks for all of the types of plants for which the restructured FutureGen was eligible to receive proposals, such as conventional pulverized-coal and oxyfuel combustion plants, but only contemplated proposals for IGCC plants;
- the risk that industry respondents might not propose an IGCC plant for the restructured FutureGen;
- the risk that industry respondents might not propose enough viable projects for the restructured FutureGen;
- the costs, benefits, and risks of making incremental changes to the original FutureGen alongside the relative costs, benefits, and risks of the restructured FutureGen; and
- any potential overlap between the restructured FutureGen and other DOE programs.

A comprehensive analysis could have supported DOE's decision making in several ways. First, it could have helped DOE assess the risk that industry respondents to DOE's request for applications under the restructured FutureGen might not propose an IGCC plant. DOE received public comments indicating that such an outcome was possible because IGCC is not yet prevalent in the industry—only two commercial IGCC plants currently operate in the United States—and other technologies may provide better opportunities to meet the restructured program's requirements, among other reasons. Applying CCS at existing, conventional pulverized coal-fired plants is important because those plants comprise almost all operating coal-fired plants in the United States and abroad. However, according to DOE, IGCC plants integrated with CCS are important for reducing CO_2 emissions in the future. Both DOE's press release announcing the restructured program and the updated draft strategic planning document, dated July 1, 2008, that DOE provided Congress indicated that the restructured program would include IGCC. The funding announcement for the restructured FutureGen highlighted the important contribution that an IGCC plant integrated with CCS would make toward the nation's energy needs, such as providing continued fuel diversity for generating electricity and mitigating dependence on more expensive and less secure sources of energy. As late as May 2008, the Secretary of Energy indicated in congressional testimony that the restructured program would likely include IGCC, stating that advances in technology and the market, in addition to regulatory uncertainty, would provide incentives for industry to begin deploying commercial-scale IGCC plants with CCS.

In addition, a comprehensive analysis could have helped DOE assess the risk that industry respondents might not propose enough viable projects from which DOE could then assess and make multiple selections. Such an analysis could also have helped DOE assess whether the new cost-share arrangement would provide sufficient incentive for enough proposals to be selective. In the draft planning documents and press release announcing the restructured program, DOE stated that it restructured FutureGen, in part, because market conditions had changed in such a way that DOE could fund multiple industry projects and accomplish even more widespread commercialization of CCS and related information sharing across the industry than what would have been accomplished by the Alliance's consortium of 11 coal producers and electric power companies. However, DOE only received a small number of applications and some proposed projects were outside the restructured FutureGen's scope. As a result, widespread commercialization and information sharing seem less likely than under the original program. DOE also asserted that the restructured program would hasten the time frame for full-scale commercial operation of CCS. However, even if DOE accepts all applicable applications, the restructured program could implement CCS sooner than the original program at only a few commercial sites rather than, as stated before, on a more widespread and international scale.

Finally, DOE also could have used a comprehensive analysis to help compare the relative costs, benefits, and risks of the restructured FutureGen with those of making incremental and other changes to the original program in order to control or offset costs. For example, prior to the decision to restructure FutureGen, FE identified and analyzed 13 options for changes to the original program, such as reducing the CO_2 capture requirement, which aimed at reducing costs while continuing to retain some of the original program's key benefits. DOE noted that some of the potential changes would have a detrimental impact on the original program's objectives, while other potential changes would not significantly impact project objectives. Of these changes, FE either recommended or noted that DOE should be willing to consider several options with potential savings from $30 million to $55 million each. Some of these scenarios were broached during negotiations with the Alliance in the fall of 2007.

CONCLUSIONS

According to DOE, electric power industry, academic, and other officials and experts, for the foreseeable future, coal, which is abundant and relatively inexpensive, will remain a significant fuel for the generation of electric power in the United States and the world. However, coal-fired power plants are also a significant source of CO_2 and other emissions responsible for climate change. Hence, for at least the near-term, any government policies that seriously address climate change will need to have a goal of significantly reducing CO_2 and other emissions from coal- fired power plants. CCS, while still in its infancy, can be a promising technology to achieve these purposes. By integrating IGCC and CCS technology at a living laboratory host facility, DOE's FutureGen program was intended to address significant technological, cost, and regulatory issues associated with the implementation of CCS at a new plant. However, in early 2008, citing concerns about a "doubling of costs," DOE abruptly canceled the original FutureGen program and announced a dramatic restructuring. The restructuring cast aside the initial concept and substituted a request for multiple projects to be proposed by industry that would retain the goal of capturing and sequestering 1 million metric tons of CO_2, and would accept technologies other than IGCC. The restructured FutureGen left open the possibility of successfully applying CCS technology to existing conventional, pulverized coal-fired power plants— an important goal in its own right, since those plants account for almost all of the coal-fired generating capacity in the United States and abroad. However, there are already existing programs to address CCS at existing plants, and the decision to remove the FutureGen program's specific focus on cutting edge technology (IGCC) at new plants was not well explained.

In at least two ways, DOE's decision, which affected potentially up to $1.3 billion in federal funding, was not well considered. First, the decision was made on the basis of a flawed comparison of life-cycle costs for the original FutureGen, in that DOE compared an estimate of constant dollars to an estimate of inflated dollars. Second, the decision was not based on a systematic and comprehensive comparison of the costs, benefits, and risks of the original FutureGen versus the restructured FutureGen. An expanded analysis of the costs, benefits, and risks of the original FutureGen compared with a range of modifications to the program could have included incremental changes to the original FutureGen program that could have preserved some of its original goals and benefits while mitigating costs. Such an analysis might also have detailed the risk that DOE would receive only a small number of applications and that those applications might not include IGCC. The analysis could also have considered whether DOE's $1.3 billion contribution for total program funding presents the best option for advancing the overall goals of CCS in both existing and future plants.

RECOMMENDATIONS FOR EXECUTIVE ACTION

To help ensure that important decisions about the FutureGen program reflect an adequate knowledge of the potential costs, benefits, and risks of viable options, and to promote the attainment of the goals of the program while protecting taxpayer interests, we are making the following two recommendations to the Secretary of Energy:

1. Before implementing significant changes to FutureGen or before obligating additional funds for such purposes, the Secretary of Energy should direct DOE staff to prepare a comprehensive analysis that compares the relative costs, benefits, and risks of a range of options that includes (1) the original FutureGen program, (2) incremental changes to the original program, and (3) the restructured FutureGen program.
2. In addition, the Secretary should consider the results of the comprehensive analysis and base any decisions that would alter the original FutureGen on the most advantageous mix of costs, benefits, and risks resulting from implementing a combination of the options that have been evaluated.

APPENDIX I. SCOPE AND METHODOLOGY

We examined (1) the goals of the original and restructured FutureGen programs, (2) the similarities and differences between the restructured FutureGen and other Department of Energy (DOE) carbon capture and storage programs, and (3) the extent to which DOE used sufficient information to support its decision to restructure the FutureGen program.

To examine the goals of the original and restructured FutureGen programs, including the results of the different approaches for meeting these goals, we reviewed relevant appropriations and agency documents, including budget justifications from fiscal years 2005 through 2009; the program plan for FutureGen that DOE submitted to Congress in 2004; the cooperative agreement between DOE and the FutureGen Industrial Alliance (Alliance), and its subsequent renewals; DOE's draft strategic planning documents and funding announcement for the restructured program; and public responses to DOE's request for information about the restructured FutureGen and its funding announcement. We also reviewed congressional testimony about FutureGen and related topics by officials from the Alliance; DOE, including the Secretary of Energy; and other knowledgeable stakeholders, such as academic and industry researchers. In addition, we met with and reviewed documents provided by officials and researchers from DOE, the Alliance, industry, nonprofit research organizations, and academia. In particular, we interviewed DOE officials from the Office of Fossil Energy's (FE) National Energy Technology Laboratory (NETL) and Office of Clean Coal. Finally, we conducted semi- structured interviews with knowledgeable stakeholders from the electric power and coal industries, nonprofit research organizations, and academia, among others. During the interviews, we discussed the goals, approaches, and anticipated results of the original and restructured FutureGen programs. Our method for conducting these interviews, including how we selected the knowledgeable stakeholders, appears in the next paragraph.

We conducted semi-structured interviews with 14 knowledgeable stakeholders from the electric power and coal industries, nonprofit research organizations, and academia, among others. We selected a nonprobability sample of stakeholders and stakeholder organizations using a "snowball sampling" technique, whereby each stakeholder we interviewed identified additional stakeholders and stakeholder organizations for us to contact.[29] Specifically, we identified the first three stakeholders to interview from previous, related GAO work and a group of contributors toward key scientific assessments of climate change and clean coal technology.[30] We then used feedback from these interviews to identify additional stakeholders to interview, and so on, being certain to interview every stakeholder or a knowledgeable official from every stakeholder organization identified by at least two other stakeholders. We also ensured that we selected stakeholders from electric power companies both within and outside the Alliance to obtain a range of industry perspectives. Over the course of our work, we conducted semi-structured interviews with knowledgeable stakeholders from the following organizations: American Electric Power, Carnegie Mellon University, the Coal Utilization Research Council, Duke Energy, Duke University, the Electric Power Research Institute, the Massachusetts Institute of Technology, the National Association of Regulatory Utility Commissioners, the National Mining Association, Resources for the Future, and Southern Company. We used a semi-structured interview guide to interview these stakeholders and facilitate analysis of what they identified as the key similarities, benefits, and differences between the original and restructured FutureGen programs, in addition to DOE’s other carbon capture and storage (CCS) programs. These semi-structured interviews allowed us to obtain information addressing all three of our objectives.

To examine the similarities and differences between the restructured FutureGen and other DOE CCS programs, we reviewed agency documents, including budget justifications from fiscal years 2005 through 2009; DOE’s Carbon Sequestration Technology Roadmap and ProgramPlan; the program plan for FutureGen that DOE submitted to Congress in 2004; DOE’s draft strategic planning documents and funding announcement for the restructured program; and relevant laws. We met with and discussed these programs with officials from NETL and FE’s Office of Clean Coal. We also conducted semi-structured interviews with knowledgeable stakeholders from the electric power and coal industries, nonprofit research organizations, and academia, among others. During these interviews, we discussed the relationship between the restructured FutureGen and DOE’s other CCS programs. Finally, we reviewed public responses to DOE’s request for information about the restructured FutureGen and DOE’s funding announcements for the restructured FutureGen and Round III of the Clean Coal Power Initiative.

To examine the extent to which DOE used sufficient information to support its decision to restructure the FutureGen program, we reviewed documents from DOE and the Alliance, including cost estmates; the cooperative agreement and subsequent updates to it; letters, presentations, and proposals documenting the renegotiation of terms for the cooperative agreement; proposed incremental changes for controlling costs; and the draft strategic planning documents and funding announcement for the restructured program. We also reviewed congressional testimony about FutureGen and related topics by officials from the Alliance; DOE, including the Secretary of Energy; and other knowledgeable stakeholders, such as academic and industry researchers. We met with and discussed the information used to support the decision to restructure FutureGen with officials from NETL and FE's Office of Clean Coal. In addition, we discussed and reviewed analyses of the costs for building coal-fired, electric power plants with officials and researchers from industry, academia, and government, including DOE's Energy Information Administration. Moreover, we discussed these costs during semi-structured interviews with knowledgeable stakeholders from the electric power and coal industries, nonprofit research organizations, and academia, among others. We also reviewed public responses to DOE's request for information about the restructured FutureGen and its funding announcement. Finally, we reviewed our recent work and guidance on best practices for cost estimation, program management, and programmatic decision making, as well as guidance from DOE and the Office of Management and Budget.

We conducted this performance audit from June 2008 to February 2009, in accordance with generally accepted government auditing standards. Those standards require that we plan and perform the audit to obtain sufficient, appropriate evidence to provide a reasonable basis for our findings and conclusions based on our audit objectives. We believe that the evidence obtained provides a reasonable basis for our findings and conclusions based on our audit objectives.

APPENDIX II. BUDGET AUTHORITY AND OBLIGATIONS FOR FUTUREGEN

Table 4. DOE Budget Authority and Obligations for FutureGen, Fiscal Years 2004 through 2008

Fiscal year	DOE budget authority for FutureGen	Adjusted DOE budgetary resources	DOE obligations for its cooperative agreement with the FutureGen Industrial Alliance	Remaining budgetary resources
2004	$9	$8.64		
2005	$18	$17.26		

2006	$18	$17.33		
2007	$54	$52.50		
2008	$75	$74.32		
Total	**$174**[a]	**$170.05**[b]	**$39.11**[c]	**$130.94**[d,e]

Source: DOE.

[a]All FutureGen budget authority was no-year authority, which means the authority is available for obligation for an indefinite time period. DOE has requested $156 million for FutureGen in its budget justification for fiscal year 2009.

[b]According to DOE, the department adjusted its program budget for several factors. In fiscal years 2004 through 2008, the program's budget authority was adjusted for the Small Business Innovation Research program and/or the Small Business TechnologyTransfer program. Moreover, the FutureGen budget for fiscal years 2004 and 2005 also included the Interior and Omnibus reduction, and in fiscal years 2004 and 2006, FutureGen's budget authority included a general rescission.

[c]According to DOE, the agency has spent over $22 million on obligations for the original FutureGen, via its cooperative agreement with the FutureGen Industrial Alliance. In addition, DOE anticipates reserving an additional $2.5 million for final close-out costs related to these obligations pending invoicing from the Alliance. Depending on the actual amount of these final close-out costs, DOE estimates that it could deobligate approximately $14.5 million of its obligations for the original FutureGen, after which time these funds would be available for the restructured FutureGen.

[d]According to DOE, the agency has additional obligations and expenditures for support contracts related to FutureGen that fall outside the scope of its cooperative agreement with the Alliance. The expenditures and remaining obligations for these contracts total approximately $7.8 million.

[e] DOE officials told us that, after accounting for all anticipated expenditures on the original FutureGen, they expect the agency to have approximately $293 million available for the restructured FutureGen in unobligated balances of unexpired budget authority, brought forward from prior years (i.e., carryover), including the $156 million budget request for fiscal year 2009.

APPENDIX III. COMMENTS FROM THE DEPARTMENT OF ENERGY

Note: GAO comments supplementing those in the report text appear at the end of this appendix.

Department of Energy
Washington, DC 20585

February 4, 2009

Mr. Mark E. Gaffigan
Director
Natural Resources and Environment Team
U.S. Government Accountability Office
441 G Street, NW, Mail 2T23A
Washington, DC 20548

Dear Mr. Gaffigan:

Thank you for the opportunity to review the Government Accountability Office (GAO) draft report entitled, "DOE's Decision to Restructure FutureGen Should Be Based on a Comprehensive Analysis of Costs, Benefits, and Risks" (GAO-09-248). Enclosed please find the U.S. Department of Energy's comments on the draft report.

If you have any questions or comments please contact Mr. Jarad Daniels of my staff at (202) 485-7355.

Sincerely,

Victor K. Der
Acting Assistant Secretary
Office of Fossil Energy

Enclosure:
DOE Comments on Draft GAO Report

See comment 1. Now on pp. 8 and 24.
See comment 2. Now on pp. 2, 17, and 18.
See comment 3. Now on pp. 3 and 26.

Department of Energy Comments on GAO "DOE's Decision to Restructure FutureGen Should be Based on a Comprehensive Analysis of Costs, Benefits, and Risks" (GAO-09-248) (GAO Draft Report)

This responds to your request for comments by the Department of Energy on the above-referenced GAO Draft Report.

1. The draft report uses inconsistent terminology to characterize the 13 options for incremental changes developed by the Office of Fossil Energy. The summary page, page 9, and the discussion on page 27 make it appear that all 13 options were viable without jeopardizing the project objective. In reality, some of the 13 options were not considered viable or recommended by Fossil Energy and were rejected by the Department as a whole.

At each of these locations (the summary page, last sentence of the third paragraph; page 9, the first full sentence; and page 27 second to last sentence of the first paragraph), DOE recommends GAO delete the number "13." Not all of the 13 options would have preserved the original project objectives nor were all recommended. Additionally on page 27, the last sentence of the first paragraph is incorrect and should either be deleted or the sentence should start with "Some of these options."

The discussion on page 33 is more accurate in that it states that DOE considered some of the potential changes within the list of 13 to have a detrimental impact on the original project objectives and pursued the remaining changes during negotiations with the Alliance. To the extent GAO believes discussion of the FE options is appropriate, DOE recommends that the report consistently present the options as they are characterized on page 33.

2. On page two, suggest that the last sentence of the carryover paragraph be changed to reflect that only Round III of the CCPI focused on carbon capture and sequestration and that selections have not yet been made. Suggested revision would read:

> *For example, DOE's Clean Coal Power Initiative (CCPI) Round III Funding Opportunity Announcement will result in cost-shared partnerships with industry for projects that demonstrate carbon capture from coal-fired plants with sequestration or beneficial reuse.*

3. On page 3, in the first full paragraph, the draft report states:

> *In December 2007, after DOE finished conducting the extensive environmental analyses required by the National Environmental Policy Act of 1969... of four potential sites that took over a year, the Alliance announced that it had selected Mattoon, Illinois, for the location of FutureGen.*

See comment 4.
Now footnote 5.
See comment 5.

In order to accurately depict the context of activities at the time, the report text should be revised. The above-quoted sentence implies that DOE completed its NEPA responsibilities prior to the Alliance announcement of the Mattoon site, and that is not the case. For projects requiring an environmental impact statement, an agency's NEPA review culminates with a Record of Decision. DOE had not issued a Record of Decision at the time the Alliance, over the Department's objection, chose to announce its site selection. The chronology of activities set out in the Cooperative Agreement clearly provided for the issuance of a Record of Decision first. This context should be provided for in the text of the GAO report.

4. On page 4, DOE recommends changing footnote 4 to read:

> *In addition, for original FutureGen, costs for the initial planning and research stages were subject to a 20 percent nonfederal cost-share.*

As drafted by GAO, the footnote sentence applies to both the original and restructured approaches, but should only apply to the original FutureGen. For Restructured FutureGen, DOE mandated not less than 50 percent private sector cost-share for all stages.

Since Fiscal Year 2005, the requirements for cost sharing of the original FutureGen has been set out in statute:

> *Provided further, That the initial planning and research stages of the FutureGen project shall include a matching requirement from non- Federal sources of at least 20 percent of the costs: Provided further, That any demonstration component of such project shall require a matching requirement from non-Federal sources of at least 50 percent of the costs of the component . . ..*

See, e.g., Pub. L. No. 108-447.

5. On pages 6 and 18, the draft report suggests that the technology transfer potential was greater under the original FutureGen because the non-profit Alliance would own new inventions and was willing to share its knowledge industry-wide. GAO contrasts the original FutureGen with the Restructured program where GAO believes for-profit recipients would be more reluctant to disseminate information. In fact, the intellectual property policies and provisions for each effort are essentially the same. For both the original and Restructured program, most new inventions would likely be generated by for-profit subcontractor/vendors supplying technology to the project. Ownership of these inventions would be granted to the subcontractors/vendors either by law for small businesses or nonprofits, or through a DOE patent waiver for large businesses. Inventions by the Alliance were far less likely since they were to have subcontracted for most technology development efforts. Furthermore, as with original FutureGen, DOE will require Restructured FutureGen recipients to disseminate the results of the project. Accordingly, the underlying assumptions for GAO's conclusions and comparisons about

See comment 6.

See comment 7. Now on pp. 3 and 26.

technology transfer potential are in part faulty. Please see DOE's comments on GAO's draft statement of facts.

In the same section on page 6, and in the first line on page 19, the draft reports states that the Alliance was comprised of utilities, mining companies, and "others." To the best of DOE's knowledge, the Alliance was comprised only of utilities and mining companies both domestic and foreign.

6. In the first line on page 8, CO_2 should have a subscript rather than a superscript.

7. On page 13 in the first paragraph, the draft report states:

> *On January 30, 2008, about a month after the Alliance announced that it had selected Mattoon, Illinois, for the site of the FutureGen plant, DOE announced that it had decided to take the FutureGen program in a different direction.*

We believe the statement set out above needs to be put in context. Prior to the announcement and without knowledge of the Alliance's choice of site, DOE asked the Alliance not to go forward with an announcement and further advised the Alliance that an announcement at that time was inadvisable in light of the ongoing DOE-Alliance discussions.

By mid-2007, in part because of an overall climate of cost escalations for building power plants around the world, DOE managers had become increasingly concerned about actual and potential escalating costs for FutureGen, and DOE had signaled to the FutureGen Alliance that DOE had serious concerns about the sustainability of the project. During September 2007, DOE began negotiating with the Alliance to develop a mutually acceptable cost-sharing formula to deal with the prospect of further cost-growth.

In December 2007, after analyzing four potential sites for the original FutureGen, the FutureGen Alliance announced that it had selected Mattoon, Illinois, for the location of the original FutureGen. The Alliance announced the site in advance to DOE's issuance of a record of decision (ROD) to complete DOE's review under the National Environmental Policy Act (NEPA) and contrary to DOE's advice. Announcement of the Alliance's selection of a site prior to the ROD was inconsistent with the schedule set out in the Cooperative Agreement.

Only after months of unsuccessful negotiations with the FutureGen Alliance regarding cost sharing for cost growth did DOE announce, on January 30, 2008, that it would not proceed beyond the then current budget period with the FutureGen Alliance, and that it planned to take a different approach to FutureGen. DOE's rationale for this decision was based on DOE's interest in maximizing the role of private sector innovation, providing a ceiling on federal contributions, and accelerating and increasing the use of clean energy technologies to help meet the growing demand for energy while also mitigating greenhouse gas emissions. The selected location of the site had no bearing on DOE's determination.

See comment 8. Now on p. 14.
See comment 9. Now on p. 15.
See comment 10.
See comment 11.
See comment 2. Now on pp. 2, 17, and 18.
See comment 12. Now on p. 23.
See comment 13. Now on p. 25.
See comment 14.
See comment 15. Now on p. 26.

8. On page 16, the first sentence of the first full paragraph states:

> *Both the original and restructured programs would operate plants at commercial size (about 300 megawatts).*

Under the restructured program, IGCC plants were expected to have a nominal capacity of 300 megawatts, but, as set out in the Funding Opportunity Announcement, non-IGCC projects needed only to be at a scale sufficient to prove commercial viability and be designed to produce and capture 1 million tons of CO_2 per year.

9. On page 17, in Table 2, under original FutureGen, the word "includes" should be deleted in the first row since the original program was exclusively focused on IGCC.

10. On page 19, DOE recommends GAO strike the words "in new inventions it would likely own" from the last sentence of the first paragraph. Inclusion of this phrase would actually limit the scope which can also more broadly include non-patentable technical data.

11. On page 21, the words "Round III" should be added after "Clean Coal Power Initiative" in the underlined heading since only the third round of CCPI has focused on carbon sequestration.

12. On page 25, change footnote "d" to read "DOE cost sharing is generally capped at 80 percent." Alternately, delete the first sentence of the footnote.

13. On page 28, DOE recommends that GAO revise the first sentence of the last paragraph to read:

> *Following this presentation, senior DOE officials directed FE to negotiate with the Alliance new cost sharing arrangements under the cooperative agreement, which was subject to a continuation decision in June 2008.*

14. On page 29, in the fourth to last sentence of the first full paragraph, DOE recommends that GAO replace the word "renewing" with "continuing" for technical accuracy.

15. On page 29, the second to last sentence of the first full paragraph, as currently drafted, erroneously suggests that DOE officials disapproved of the selection of Mattoon for the project site. The sentence implies that DOE's determination was influenced by the Alliance's selection of the Mattoon site, and that is not the case. Prior to the announcement and without knowledge of the Alliance's choice of site, DOE asked the

See comment 16. Now on p. 26.

See comment 17. Now on p. 29.

See comment 18. Now on pp. 28, 29, and 38.

See comment 19. Now on p. 38.

announcement was inadvisable. The selected location of the site had no bearing on DOE's determination to restructure the program.

DOE recommends that the second to last sentence of the first full paragraph be revised into two sentences as follows:

> *DOE officials also stated that they disapproved of the Alliance's decision to announce the selection of a project site. Prior to the announcement and without knowledge of the Alliance's choice of site, DOE asked the Alliance not to go forward with an announcement and further advised the Alliance that an announcement was inadvisable. DOE had not issued its NEPA Record of Decision and was continuing its attempts to negotiate with the Alliance.*

16. On page 29, the sentence at the end of the first paragraph should be modified to make it clear that DOE lost confidence in the ability of the Alliance to fund its share of the project cost, rather than confidence in the Alliance member companies or their representatives.

17. On page 32, the third to last sentence of the first paragraph suggests that the draft strategic planning document was developed in July 2008, and that is not correct. The document was transmitted to Congress in response to a document request in July of 2008. In addition, we would also like to clarify a point regarding DOE's position on the role of CCS, including IGCC with CCS, in reducing CO_2 in the future. Therefore the sentence should be changed to read as follows:

> *However, according to DOE, plants with CCS, including those with IGCC integrated with CCS, are important options for reducing CO_2 emissions in the future, and DOE's press release announcing the restructured program and the draft strategic planning document that DOE provided to Congress in July 2008 indicated that the restructured program would include IGCC.*

18. On page 32, footnote 30 and on page 40, table II.2 footnotes b and c, the draft report states: "DOE had not yet provided" the information or response that GAO had requested. As of the date of these comments, DOE understands that it has either already provided the relevant documentation, and has referred GAO back to those documents, or has since provided the documentation. DOE understands that all information has been provided.

19. On page 40, we noticed several errors in the Tables contained in Appendix II.
Table II.1

The correct FutureGen appropriations for Fiscal Year 2004 through 2008 are as follows:

2004 $9,000,000
2005 $18,000,000
2006 $18,000,000

Total $174,000,000

These above-stated amounts are not yet adjusted for rescissions or the SBIR/STTR offset. Hence, the actual amount available to DOE for FutureGen, as depicted in Table II.2, is less than $174,000,000. Under the "Appropriated" column in Table II.1, the Fiscal Year 2007 amount should be changed from $18,000,000 to $54,000,000. The total should be changed accordingly. Under the heading "Directed" in Table II.1, the Fiscal Year 2007 amount of $257,000,000 should be deleted as this was deferred budget authority under the Clean Coal Technology Program that was not made available to the Department for FutureGen. The last column should be updated to reflect the changes in the first two columns.

Table II.2

The calculations for the Fiscal Year 2006 Budget are as follows:

Appropriation Act	$18,000,000
General Rescission 1%	(180,000)
SBIR Assessment	(441,000)
STTR Assessment	(53,000)
FY-2006 Available	$17,326,000

The Fiscal Year 2006 amount in Table II.2 should be changed from $17,820,000 to $17,326,000 and the Total changed to $170,045,000.

The following are GAO's comments on the Department of Energy's letter dated February 4, 2009.

GAO's Comments

1. We modified our report to address DOE's concerns about our discussion of the 13 options for incremental changes.
2. We modified our report to add clarifying information on Round III of the Clean Coal Power Initiative.
3. We added clarifying information about the timing of the site selection announcement and the release of DOE's NEPA Record of Decision.
4. We revised the footnote to state that, according to DOE, not less than a 50 percent nonfederal cost share will be required for all of the restructured FutureGen's stages.

5. We have revised the report to remove the referenced discussion.
6. We made DOE's editorial correction.
7. The report does not state or imply that the location of the site was the reason for the program's restructuring, but rather states that DOE's restructuring decision was based on a desire to contain costs in a time of increasing cost pressures. However, we revised the report to clarify that the Alliance announced its site selection decision before DOE's Record of Decision was released—which has not happened, as of the date of this report.
8. DOE clarifies that under the restructured program, IGCC plants are expected to have a nominal capacity of 300 megawatts, but non-IGCC projects need only be at a scale sufficient to prove commercial viability and be designed to produce and capture 1 million tons of CO_2 per year. We revised the report to reflect this information.
9. As suggested by DOE, we revised table 2 to reflect that the original FutureGen program focused exclusively on IGCC.
10. As indicated in our response to comment 5, we have revised the report to remove the referenced text.
11. We made appropriate revisions to the report to reflect that Round III of the Clean Coal Power Initiative focuses on carbon sequestration.
12. We revised table note "d" to table 3, to state that DOE's cost sharing is generally capped at 80 percent.
13. We revised the report to clarify that senior DOE officials directed FE to negotiate new cost sharing arrangements under the cooperative agreement with the Alliance.
14. We revised the report to use the word "continue" in place of "renew" wherever it would more accurately reflect the various stages of the cooperative agreement.
15. Our draft report did not insinuate or state that DOE did not favor the Mattoon, Illinois, site or that DOE's restructuring decision was based on a disapproval of the Alliance's site selection announcement. Rather, our report states that DOE's decision was based on a desire to limit its exposure to increased costs. However, as suggested by DOE, we clarified the report by adding that DOE had instructed the Alliance to not announce the site selection before DOE could release the Record of Decision.
16. We agree with DOE and revised the report to clarify that DOE lost confidence in the ability of the Alliance to fund its share of the project

cost, rather than that DOE lost confidence in the Alliance members or their representatives.

17. We revised the report to clarify that both DOE's press release announcing the restructured program and the updated draft strategic document, dated July 1, 2008, that DOE provided Congress indicated that the restructured program would include IGCC.
18. DOE provided information regarding the official basis for restructuring the FutureGen program and its budget authority, obligations, and expenditures that we incorporated into our report, including table 4 and its table notes. We also included in the report an additional assessment of documents, to which DOE referred as providing the basis for its decision to restructure FutureGen.
19. Regarding the tables in appendix II of the draft report, DOE provided updated FutureGen appropriations information for fiscal years 2004 and 2007 and certain calculations for the fiscal year 2006 FutureGen budget. In response, we merged tables 4 and 5 from the draft to create one table in the final report, and we adjusted the figures and calculations for the data that DOE provided.

End Notes

[1] Currently, only two IGCC plants operate at commercial scale in the United States. In service since 1997, the Polk Station, near Mulberry, Florida, can provide 250 megawatts to the electric grid. The Wabash River Coal Gasification Repowering Project is the first full- size commercial gasification-combined cycle plant built in the United States, having begun operations in November 1995. The plant, located outside West Terre Haute, Indiana, can provide 262 megawatts to the electric grid.

[2] Pub.L. No. 9 1-190 (1970). Under the act, federal agencies must evaluate the likely environmental effects of their activities using an environmental assessment or, if the activity likely would significantly affect the environment, a more detailed environmental impact statement.

[3] The cooperative agreement originally scheduled the final site announcement to take place on September 4, 2007, contingent upon DOE's August 1, 2007, publication of the Record of Decision. On November 9, 2007, DOE issued a press release announcing the completion of its final environmental impact statement for FutureGen and that DOE anticipated site selection would be made later that year.

[4] Pub. L. No. 109-58 (2005), *codified at* 42 U.S.C. § 16352.

[5] In addition, beginning in fiscal year 2005, costs for the initial planning and research stages were subject to a 20 percent nonfederal cost share. According to DOE, for the restructured FutureGen, not less than a 50 percent nonfederal cost-share for all stages will be required.

[6] DOE has identified certain details regarding the negotiations for both the original and the restructured FutureGen as sensitive or proprietary information. Due to the ongoing nature of these negotiations for the restructured FutureGen and the fact that disclosure of sensitive/proprietary information could adversely affect negotiations of these projects and

related future projects, our discussion of some aspects of these negotiations is necessarily general.

[7] GAO, *Defense Acquisition: Termination Costs Are Generally Not a Compelling Reason to Continue Programs or Contracts That Otherwise Warrant Ending*, GAO-08-379 (Washington, D.C.: Mar. 14, 2008); *Cost Assessment Guide: Best Practices for Estimating and Managing Program Costs—Exposure Draft*, GAO-07-1134SP (Washington, D.C.: July 2, 2007); *Standards for Internal Control in the Federal Government*, GAO/AIMD-00-21.3.1 (Washington, D.C.: Nov. 1, 1999); and *Executive Guide: Leading Practices in Capital Decision-Making*, GAO/AIMD-99-32 (Washington, D.C.: Dec. 1, 1998).

[8] The International Energy Agency identifies CCS and other clean coal technologies as one of the most promising routes for mitigating emissions and notes that, "CCS could reconcile continued coal burning with the need to cut emissions in the longer term." Similarly, the IPCC notes that CCS would help preserve existing energy infrastructure, thereby restraining the cost of emissions reductions.

[9] Regardless of the approach used to generate electricity and capture CO_2, additional electricity, often referred to as the energy penalty or parasitic power, is required for capture and compression.

[10] ICGG plants can be either air-blown or oxygen-blown, referring to the way in which oxygen is introduced into the gasifier. The syngas produced in an air-blown gasifier consists primarily of inert nitrogen, along with hydrogen, carbon monoxide, low levels of CO_2, and water vapor. In oxygen-blown gasification, the syngas is comprised of the same primary components, but has very little nitrogen and is, therefore, more concentrated in respect to the other components.

[11] GAO, *Climate Change: Federal Actions Will Greatly Affect the Viability of Carbon Capture and Storage As a Key Mitigation Option,* GAO-08-1080 (Washington, D.C.: Sept. 30, 2008).

[12] The Alliance is currently comprised of the following 11 companies that produce coal or generate coal-fueled power on six continents: American Electric Power, Anglo American LLC, BHP Billiton, China Huaneng Group, CONSOL Energy Inc., E.ON U.S., Foundation Coal Corporation, Peabody Energy, Rio Tinto Energy America, Southern Company, and Xstrata Coal Pty Limited. The Alliance membership continues to be open to U.S. and international companies that produce coal or electricity from coal as a significant business activity.

[13] In 2006 and 2007, DOE received funding contributions for FutureGen from the governments of India and South Korea.

[14] According to DOE, if a non-IGCC plant is selected under the restructured FutureGen, it is only required to be at a scale sufficient to prove commercial viability and be designed to produce and capture 1 million tons of CO_2 per year. IGCC plants would be required to produce at least 300 megawatts of gross electricity output.

[15] GAO-08-1080.

[16] While the first two rounds of CCPI did not focus on CCS, Round III does so through projects that capture and store CO_2 or put CO_2 to beneficial reuse. Round I was broadly focused on advancing technologies in coal-fired power generation that would result in efficiency, as well as environmental and economic improvements. Round II was focused on gasification technology and mercury control.

[17] According to DOE, most oil is produced in three distinct phases: primary, secondary, and tertiary, or enhanced oil recovery (EOR). The definition of tertiary or EOR is that a substance, such as CO_2, is added to the reservoir after secondary recovery in order to increase production. The purpose of EOR is to increase oil production, primarily through an increase in temperature, pressure, or an enhancement of the oil's ability to flow through the reservoir.

[18] We reported in GAO-08- 1080 that the regional partnerships program appears to be placing more emphasis on demonstrations of CO_2 capture at coal-fired power plants. Specifically, a

DOE official identified three projects being planned to capture CO_2 from coal-fired power plants, including possibly capturing 500,000 metric tons of CO_2 from a coal-fired power plant in North Dakota. Program shifts were also evident in the Innovations for Existing Plants and CCPI programs. We recommended that DOE continue its recent budgetary practice of helping to ensure that greater emphasis is placed on supporting technologies that can reduce greenhouse gas emissions at existing coal-fired power plants.

[19] Phase I of the regional partnerships, the Characterization Phase (2003-2005), focused on describing the potential for CO_2 storage in deep oil-, gas-, coal-, and saline-bearing formations. Phase II, the Validation Phase (2005-2009), is implementing 25 small-scale geologic storage tests. Phase III, the Deployment Phase (2008-2017), is a continuation of the Phase II small-scale tests, but at a much larger scale.

[20] In response to language in the Explanatory Statement accompanying its fiscal year 2008 appropriation, DOE has shifted the focus of the Innovations for Existing Plants program to research and development on CO_2 capture technologies that can be retrofitted to existing pulverized coal-fired power plants.

[21] Federal loan guarantee programs help borrowers obtain access to credit with more favorable terms than they may otherwise obtain in private lending markets because the federal government guarantees to pay lenders if the borrowers default, which makes extending credit more attractive to lenders.

[22] GAO, *Department of Energy: New Loan Guarantee Program Should Complete Activities Necessary for Effective and Accountable Program Management*, GAO-08-750 (Washington, D.C.: July 7, 2008).

[23] The Alliance estimate was $ 1.785 billion, in current dollars, from 2005 through 2017.

[24] We selected constant fiscal year 2005 dollars for illustrative purposes. However, our review of DOE's documentation pertaining to its $950 million cost estimate for the original FutureGen shows that DOE considered base year constant dollars for the estimate from several years, ranging from late quarter 2003 dollars, to fiscal year 2004 dollars, and fiscal year 2005 dollars. We asked DOE to confirm the base year for its cost estimate, and department officials responded that several versions of the estimate had been prepared using fiscal year 2003 and fiscal year 2004 dollars. When we pointed out the existence of an additional estimate in fiscal year 2005 dollars, DOE officials informed us that they would clarify which constant year dollars DOE had used. However, as of January, 2009, DOE had not yet fully clarified this information for us.

[25] The Alliance has decided to continue preparing the preliminary cost estimate, and it anticipates completing and releasing the estimate by early 2009.

[26] The cooperative agreement also stated that DOE and the Alliance agreed in principle to, ultimately, cap DOE's share of costs at $700 million in constant 2004 dollars—approximately 74 percent of DOE's cost estimate—and it required the Alliance to develop, by June 2008, a proposal for the terms of such a cap. The cap was to be adjusted for unanticipated cost escalation upon each scheduled renewal or continuation of the cooperative agreement, based on a suitable index of actual costs negotiated by DOE and the Alliance.

[27] U.S. Senate. Committee on Appropriations, Testimony of Samuel W. Bodman, Secretary of Energy, U.S. Department of Energy, before the Subcommittee on Energy and Water Development, May 8, 2008.

[28] These findings are for building power plants, in general, and do not specifically address FutureGen.

[29] The information gathered from these semi-structured interviews cannot be used to generalize findings to, or make inferences about, the entire population of knowledgeable stakeholders on FutureGen and clean coal technology. Although the sample provides some variety, it is unlikely to capture the full variability of knowledgeable stakeholders and it cannot provide comprehensive insight into the views of any one group of knowledgeable stakeholders. This is because, in a nonprobability sample, some elements of the population being interviewed

have no chance, or an unknown chance, of being selected as part of the sample. However, the information gathered during these semi-structured interviews allows us to discuss various stakeholder views on FutureGen and clean coal technology, and it provides important context overall. It also helps us interpret the documentation and testimonial evidence we have collected.

[30] Massachusetts Institute of Technology, *The Future of Coal: Options for a Carbon-Constrained World* (Cambridge, MA, 2007) and IPCC, *IPCC Special Report on Carbon Dioxide Capture and Storage* (Montreal, Canada, Sept. 2005).

CHAPTER SOURCES

The following chapters have been previously published:

Chapter 1 – This is an edited, excerpted and augmented edition of a United StatesCongressional Research Service publication, Report Order Code RL34621, dated April 15, 2010.

Chapter 2 – This is an edited, excerpted and augmented edition of a United StatesGovernment Accountability Office publication, Report Order Code GAO-10-675, dated June 2010.

Chapter 3 – This is an edited, excerpted and augmented edition of a United StatesGovernment Accountability Office publication, Report Order Code GAO-09-248, dated February 2009.

INDEX

D

E

R

S

T

U

V

W